3 8027 00008240 3
AF443161

Guidelines for assuring testability

Published by: The Institution of Electrical Engineers, London

© 1988: Institution of Electrical Engineers

While the authors and the publishers believe that the
information and guidance given in this work is correct, all
parties must rely upon their own skill and judgment when
making use of it. Neither the authors nor the publishers
assume any liability to anyone for any loss or damage
caused by any error or omission in the work, whether such
error or omission is the result of negligence or any other
cause. Any and all such liability is disclaimed.

ISBN 0 86341 129 0

Contents

GENERAL PRINCIPLES

Section One: Introduction to the Guidelines

Section Two: Management for Testability

Section Three: Assuring Testability

TECHNICAL CONSIDERATIONS

Section Four: Software Testing and Testability

Section Five: Electronic Devices and Assemblies

Section Six: Electrical and Electromechanical Products and Complex Systems

DESIGN RULES

Section Seven: Testability Design Rules

APPENDICES

APPENDIX A : Glossary

APPENDIX B : Some Available Tools

APPENDIX C : Standards

APPENDIX D : Bibliography

Acknowledgements

These Guidelines have been prepared by a Working Party of the Computing Standards Sub-Committee within the Computing and Control Division of the Institution of Electrical Engineers, and edited by Mr C D Marsh of the Institution's Technical Affairs Department. The work has been assisted by the Department of Trade and Industry under the Support for Innovation Scheme.

The Working Party consisted of the following members:

Mr D P B Prosser	(Independent Consultant) Chairman
Mr D J Allan	(GEC Power Transformers)
Dr A P Ambler	(Brunel University)
Mr C F Amor	representing the Design Council
Dr R G Bennetts	(Consultant)
Dr L J Bond	(University College London)
Mr R S Dunn	(MoD Directorate of Standardization)
Mr R J Earl	(Racal Marine Electronics)
Mr P J B Fergus	(Consultant)
Mr P J Giles	(SHL Reliability)
Mr D J Gray	(APV Automation)
Mr A Hann	(Consultant) representing the Institution of Electronic and Radio Engineers
Brigadier R Knowles	representing the Institute of Quality Assurance
Mr D I Mackinnon	(Department of Trade and Industry)
Mr C M Maunder	(British Telecom Technology Executive)
Mr R Metcalfe	representing the Institution of Mechanical Engineers
Mr D Murray	(Consultant)
Mr M J Pilditch	(Computer Sciences Company)
Mr J B Stewart	(British Steel Corporation)
Mr P J Tullett	(Plessey Naval Systems) representing the Electronic Engineering Association
Dr K R Whittington	(TI Research) representing the British National Committee for Non-Destructive Testing

Further valuable advice was obtained from the members of a large Consultative Committee set up to represent a broad spectrum of experience across the electrical engineering industry.

The IEE is grateful for the contributions of time and expertise from all who have assisted with the production of these Guidelines.

Figures 1 and 2 on pages 5-9 and 5-10 are © C M Maunder and are reproduced with permission.

Executive Summary

Testing is an essential activity

> in **design**, to establish performance and check conformance to specification

> in **manufacture**, to prove correct transfer of the design and eliminate process defects

> in **service**, to check continuing quality of operation, rapidly locate faults and facilitate repair.

With the advent of software and micro-chip technology pervading all branches of engineering, the cost and difficulty of testing is increasing rapidly due to

> the problems of testing software

> reduction in the physical size of components, sub-assemblies and equipments, making test access difficult

> increase in product complexity.

High quality and reliability are essential attributes of any successful product in today's fiercely competitive market environment. Testing plays a crucial role in achieving these objectives. The requirements for testing during all stages of design, manufacture and in-service life must be included as part of the initial design specification and product costing exercise.

Testability must be an essential objective of the management and design process, to avoid designing, manufacturing and marketing products which are extremely difficult to analyse during design, to test in production and which are unreliable and costly to maintain in service.

Products can only be thoroughly tested if they are designed to be testable right from the initial concept—that is the clear message of these Guidelines.

It is now often necessary to include hardware or software Built-in-Test (BIT) facilities into the design of a product in order to make it testable and maintainable to the required degree. The resulting increase in design complexity is then outweighed by the savings in ease of production and maintenance.

In order to achieve the benefit of a major reduction in life cycle costs and an increase in customer satisfaction, management must take on the responsibility for ensuring that

> testability objectives are included in design requirement specifications

> manufacturing and maintenance test strategies for the product are produced

> a testability assurance programme is agreed

> testability reviews are carried out during design and production phases, for example, in Design Review and Design Audit

testability features are not traded out under pressure for meeting time targets, minimising design cost or maximising performance, except by a conscious managerial decision

a documentation system is in place for ensuring feedback from manufacturing and field maintenance activities to the design area

life cycle costs are *always* kept firmly in mind.

The Guidelines are intended to give guidance to general management, product and project managers as well as to designers. General sections on testability assurance are followed by sections relating more specifically to testability in software, electronics and electromechanical products and systems. The final section contains a list of some appropriate rules.

The appendices contain a list of available testability tools, a Glossary and a Bibliography for those readers wishing to follow up topics in greater depth.

General Principles

SECTION ONE : Introduction to the Guidelines

SECTION TWO : Management for Testability

SECTION THREE : Assuring Testability

Section One: Introduction to the Guidelines

1.1 General

Who gains from testable products?

The manufacturer—because his products are easier to test and trouble shoot, resulting in increased throughput and more efficient use of test equipment, test programming and manufacturing resources.

The marketing manager—because his products have additional customer appeal, through higher quality, faster servicing and lower support, down-time, inventory and test equipment costs.

The customer—because the equipment he buys is easier and less costly to support with less time wasted in having equipment that is not available to perform its designated function.

The customer support group—because of lower warranty charges and less costly logistical support.

Testing accounts for a very substantial percentage of the whole life cost of engineering products, a percentage which is increasing as product complexity increases; for example, a figure of 30%, or even higher, is not uncommon for the testing costs of a complex electronic system. **Testing is therefore a prime cost-reduction target for managers, designers, manufacturers and users.** Consequently, whilst these Guildelines are aimed primarily at product and project managers, it is important that general managers, designers, production engineers and the ultimate users recognise the importance of the concept of testability and participate in the development of cost effective programmes for its achievement. Such testability assurance programmes should take account of activities from the initial concept of a product and the specification of requirements right through the feasibility study, design, development, component procurement, production, commissioning and maintenance phases.

Testability assurance programmes can have a significant effect in reducing life-cycle costs. Their implementation involves the management of design, production, installation, commissioning and in-service support activities and, as a result, management responsibilities may need to be reviewed. **Managers, at all levels, are recommended to study these Guildelines and consider how testability assurance programmes affect their activities and responsibilities.**

1.2 Scope and Structure of the Guidelines

These Guidelines deal with testability, the attribute which enables the designer, the manufacturer and the user to establish and verify the defined characteristics of engineering entities in a cost-effective way.

Section One deals with general principles, concepts and philosophies.

Section Two covers the management aspects of introducing testability assurance programmes as part of an overall product development strategy.

Section Three contains basic definitions and covers the application of checklists as an aid to assessing the testability of a design.

Sections Four to Six deal with special considerations relating to a number of different areas of engineering.

Section Seven contains checklists of points that need to be covered in design reviews and audits in order to achieve a testable product.

Using these Guidelines

All managers, whether in design, manufacturing, maintenance or general management, are recommended to study Sections One, Two and Three and consider the implications for their areas of responsibility. In addition, other Sections may prove useful in providing a background of awareness of the more detailed aspects of designing and achieving cost-effective testable systems.

Product and project managers should be concerned with the study and application of all Sections of the Guidelines, as they affect each phase of the life-cycle of a project. In particular, Section Seven contains a series of detailed checklists for guidance to designers, and for the use of project managers and quality assurance personnel during design reviews and audits.

Designers need to study not only the complete text of the Guidelines, but also to follow up by reading some of the more detailed texts included in the Bibliography.

Students at universities and polytechnics will find the Guidelines to be a useful introduction to the subject of designing testable components, equipments and systems.

1.3 General Philosophies

It is essential that testability requirements are evolved and specified as part of the performance and engineering specifications, but **before effective criteria for testability can be established, production and maintenance strategies must be determined and harmonised.**

The key aspects of the application of the concept of testability can be summarised as follows:

(a) Testability is the inherent property of an entity to facilitate the verification of its specified performance by defined techniques;

(b) Testability should satisfy both performance and diagnostic test requirements and allow a failed area to be identified and isolated with a defined degree of confidence;

(c) Testability is a key element of design, production, commissioning and maintenance. **Therefore management control of testability is essential.**

1.4 General Testability Requirements

The essential requirements for effective testability are:

(a) Each entity should have defined attributes which are capable of verification to the depth required by the design philosophy and the production and maintenance strategies;

(b)   Test access to each entity should be sufficient to allow complete performance testing and assessment in line with the production and maintenance strategies with the minimum disturbance to the permanent structure of the entity;

(c)   Each entity should be designed so that its performance can be verified by simple effective test methods which have a minimal effect on the operational life;

(d)   In the case of software, each entity should be documented to make the intended functionality visible, so that its effect can be verified by exercising it, and its performance determined without interfering with the internals. Planning these tests in advance is the way of ensuring testability.

1.5 Testability during Development, Manufacture and Operational Life

Testing and testability requirements for development and manufacture will differ considerably from those required during service. Test requirements during production usually involve stage by stage testing as the system is built up from its individual components during manufacture and assembly, possibly with internal adjustment and alignment of assemblies. During servicing, however, testing starts at the system level and works down to the individual components. The user also requires test results to indicate the state of operational fitness or performance deterioration at any time and a diagnostic capability which indicates any sources of malfunction or deterioration and relates these to a replaceable part or an adjustment process. **Such in-service requirements should be included as part of the initial design concept and specification.**

1.6 Trade-offs and Costs

Any reduction in achievable testability is likely to lead to a significant increase in the life cycle costs for the entity and system. **Changes in the testability specification should not be made without careful consideration of the consequences on overall life cycle costs.**

In an ideal situation, the supplier can provide a system which satisfies a customer's requirements in terms of performance, reliability and cost of ownership. In reality, however, the constraints of time and cost usually result in a system in which some product attributes have to be regarded as of secondary importance. Unfortunately, testability often falls into this category and is traded-off against what are thought to be more commercially or operationally significant attributes. **If a design trade-off leads to a low testability rating, this decision should be highlighted and the consequences pointed out to both supplier and user management.**

1.7 Organising for Testability

In order to establish and maintain an effective testability assurance programme, testability management must form a close liaison with all design disciplines.

The testability manager must be prepared to work constructively with design, manufacturing and maintenance engineers to ensure a proper balance between performance, cost and supportability. It is not efficient or effective for the testability manager to assume the role of a post-design critic and risk large cost increases and programme time slippages.

The testability assurance programme must operate from the initiation of the design effort, using design guidelines, design rules and rule checking, design reviews and audits, training programmes and objective measures to aim for "right first time" testable designs.

1.8 Product Liability

The European Commission Directive on product liability (85/374/EEC) was adopted on 25 July 1985 after some nine years of discussion. The Directive has been implemented in the UK by Part I of the Consumer Protection Act 1987, which came into force on 1 March 1988. This imposes strict liability on the producer for personal injury and damage to private property which arises from a defective product, irrespective of any fault or negligence on the part of the producer.

The advent of strict product liability legislation in the USA has, in recent years, caused designers and manufacturers there to give urgent attention to the adequate testing of their products, particularly in view of the heavy penalties which have been awarded in the Courts when judgements have been made against manufacturers of defective products.

It is not expected that the application of the Consumer Protection Act in the UK will lead to the excessive levels of damages and the associated escalation in insurance premiums seen in the USA. This is partly because damages in the UK are set by a judge rather than by a jury and partly because there is no "contingency fee" system of payment to lawyers such as has led to widespread aggressive litigations in the USA. However, it will be clearly in the interests of producers to ensure that their products are thoroughly tested. Products can only be thoroughly tested if they are designed to be testable— that is the message of these guidelines. Moreover, **complete documentation of the testing strategy, testability reviews and test results needs to be maintained and archived for the life of the product, in case it needs to be produced as evidence.**

1.9 Product Evolution

Whereas, in some areas of rapidly developing technology, products may become obsolescent in a very short time, there are still many industries in which systems and their equipments maintain a long life span of useful service. During the period of this long service use, on-site updates, enhancements, refurbishments or rebuilds are likely to take place. It is essential that at each of these events, the opportunity is taken of reviewing the total on-site testability and diagnostic capability and requirement. The cost effectiveness of adopting more up-to-date test techniques and equipment throughout the plant or equipment should be considered.

In designing new equipment for incorporation into an existing installation, on the other hand, it may be that undesired restrictions on the methods of test may be incurred in order to match the existing maintenance strategy of the installation. In such a case the designer should anticipate the possibility of later changes which would allow the use of more cost-effective test techniques.

1.10 Product Quality

High standards of quality and reliability are essential attributes of any product if it is to be successful in a highly competitive market environment. Testability plays a crucial role in achieving these objectives.

Because quality cannot be tested into a product during manufacture, designing for testability is a basic requirement.

Quality assurance systems such as BS5750 and NATO AQAP-1 should be implemented by all organisations concerned with designing and producing a successful product. The use of such systems ensures the establishment of an environment having designated management responsibilities, job descriptions, practical standards, working instructions and procedures for each area of operation. Testing must be an essential part of the programmes for exercising control over design, development, manufacturing and commissioning. Good **documentation** and an effective **change control system** are essential features of quality assurance.

Section Two: Management for Testability

2.1 Introduction

Inadequate testability in a product has an adverse effect on life cycle costs, on the economics of manufacture and in-service operation and inevitably on customer satisfaction. Consequently, **the provision of adequate testability is a matter of prime concern for the top management of both supplier and customer.**

This Section covers the following topics:

2.2 Customer involvement and responsibilities
2.3 Testability assurance programmes
2.4 Standardisation
2.5 Data collection, analysis and documentation
2.6 Partitioning and subsystem control
2.7 Condition monitoring
2.8 Safety considerations
2.9 Test economics
2.10 Training
2.11 Use of expert systems

2.2 Customer Involvement and Responsibilities

There can be two different types of "customer" for a product:

(a) where a customer or group of customers commission a supplier to design and produce an engineering entity for them;

(b) where a manufacturer designs and produces an engineering entity for a perceived market, in which case there are many different customers for the same or similar products.

In the case of (a) above, the customer should be deeply involved with the designer and manufacturer in the specification and achievement of testability. This involvement continues from the initial specification of requirements and agreement of the production and maintenance philosophies through the whole life cycle of the product covering design, development, production, commissioning and acceptance to in-service operation and maintenance.

In the case of (b) above, there is no single customer and the requirements for a 'target customer' are usually set by the Marketing Department. It is essential that the designer/manufacturer of the products should appoint a person who can perform those functions of testability management which are performed by the customer in the customer/contractor relationship in (a) above. This appointment will normally be from within the manufacturer's own organisation, possibly selected from the Marketing or Quality Assurance departments. **However, it is necessary for the person appointed to have sufficient authority to negotiate the testability assurance programme effectively** and take account of any trade-offs so that overall cost and quality objectives are achieved.

2.3 Testability Assurance Programmes

2.3.1 Scope of Testability Assurance Programmes

This guide recommends an integrated approach to the planning of testability assurance programmes.

A testability assurance programme will include:

Specification of testability requirements

Design-for-test techniques and procedures

Prediction of test coverage

Demonstration of testability achievement

Documentation

Testability review (as part of Design Review or Audit)

Prediction of test quality

Data collection, analysis and evaluation

Corrective action.

This guide also recommends the integration of the testability assurance programme with other closely related programmes for design, quality assurance, production engineering, reliability assurance, maintenance and support.

2.3.2 Testability Assurance Programme Requirements

A testability assurance programme needs to cover the following:

Preparation of a testability assurance programme plan

Establishment of sufficient, achievable and affordable testability, built-in and external test requirements and facilities to meet the agreed production and maintenance strategies

Integration of testability, for both manufacture and maintenance, into equipments and systems during the design process

Evaluation of the extent to which a design and its production meets testability requirements

Inclusion of testability in the overall programme review process

Documentation of testability reviews and findings

Documentation of field maintenance reviews and findings.

2.3.3 Application of Requirements

It should be understood that detailed requirements have to be selectively applied and tailored as needed and as appropriate to particular systems and equipment acquisition programmes.

The rationale and guidance for the selection and tailoring of testability assurance programmes can be found in MIL-STD-2165 (Appendix A). This is recommended reading for anyone wishing to perform these tasks thoroughly, although it may be found difficult reading. Useful guidance may also be obtained from DEF STAN 00-13/2.

2.3.4 Testability Assurance Programmes in Perspective

The selection of tasks for a testability assurance programme which can meet the defined testability requirements is a difficult problem for supplier and customer faced with the usual severe constraints on funding and timing.

The following sections aim to provide assistance in selecting tasks for a test programme. Once appropriate tasks have been selected, each task must be tailored in terms of timing, comprehensiveness and end products to meet overall product programme requirements.

The planned testability assurance programme must be an integral part of the overall systems engineering process and serve as an important link between design, production, commissioning and in-service operational maintenance.

2.3.5 Programme phases

The testability assurance programme plan has to be integrated with all the other activities of product development and exploitation and through all phases, including:

(a) **Feasibility Study (FS)**—the study period during which the identification and exploration of alternative solutions or solution concepts to satisfy a validated need take place.

Activities

> Identify manpower, logistic, reliability, maintainability and testability objectives critical to the system requirements and support costs
>
> Establish availability of test tools, equipment and skills
>
> Estimate what is achievable and affordable for each testability objective.

(b) **Project Definition (PD)**—the period during which selected candidate solutions are refined through extensive study and analyses; hardware and software development if appropriate; test and evaluation.

Activities

> Conduct trade-offs between system design characteristics and support concepts and costs to minimum life cycle costs
>
> Establish a consistent set of goals and threshholds for testability, reliability, maintainability, availability, BIT, manpower and logistic parameters
>
> Establish Test and Evaluation Plans to assess the achievement of the above threshholds.

(c) **Full Scale Development (FSD)**

Activities

> Perform detailed analysis and trade-offs of design, manufacturability, testability, maintainability, reliability, etc.
>
> Carry out Design Reviews or Audits, including testability reviews
>
> Test and evaluate the design against the specified concepts and objectives.

(d) **Production and Deployment Phase**

Activities

> Evaluate the plans for maintenance; support operation, support costs and manpower requirements
>
> Correct deficiencies.

2.3.6 Levels of Complexity and Associated Tasks

It is convenient to consider three levels of complexity when developing products:

Systems

Equipment

Items

The testability assurance programmes are related to these three levels. For each level, a set of testability assurance tasks needs to be prepared.

(a) **Systems**

For systems the testability tasks for each programme phase are listed below and summarised in Table 1.

Feasibility Study:

Establish testability objectives in preliminary systems specification.

Project Definition:

Prepare testability assurance programme

Incorporate testability features into items and evaluate effectiveness

Select alternative system concepts. Establish testability requirements in system specification. Allocate testability requirements to item development specifications

Conduct testability review (as part of system design review).

Full Scale Development:

Prepare and revise testability assurance programme

Incorporate testability features into all items of the system and evaluate effectiveness

Demonstrate system testability effectiveness.

Production and Deployment:

Collect data on achieved testability effectiveness. Take corrective action as required.

TASK	PROGRAMME PHASE			
	Feasibility Study	Project Definition	Full Scale Development	Production & Deployment
Testability assurance programme planning		G	G	
Testability reviews	G*	G	G	S
Testability data collection & analysis planning		S	G	G
Testability requirements	G*	G	G	
Testability preliminary design analysis		S	G	S
Testability detail design & analysis		S	G	S
Testability demonstration		S	G	S

KEY: G — Generally applicable
S — Selectively applicable to high risk items during Project Definition
    or to design changes during Production & Deployment phases
*   See also MIL-STD-1388-1 Logistic Support Analysis

Table 1 Testability Tasks for Systems

(b) **Equipment**

The following testability tasks are necessary:

Establish equipment or sub-system testability requirements with user and obtain formal agreement

Prepare testability assurance programme

Incorporate testability features into items and evaluate effectiveness

Collect data on achieved testability effectiveness.

(c) **Items**

The following testability tasks are generally needed:

Preliminary design phase:

Prepare testability assurance programme (if a programme has not been developed as part of an overall system of which the item is a part)

Incorporate testability features into the preliminary design

Prepare testability checklist for each item

Conduct testability review as part of design review.

Detail design phase:

Incorporate testability features and predict item testability:

Predict test effectiveness

Conduct testability review as part of critical design review

Demonstrate testability of item.

2.3.7 Iterations

Certain tasks are highly iterative in nature and will recur at various times during the life cycle of the system, proceeding to lower levels and greater detail in the usual systems engineering manner.

2.4 Standardisation

In many cases utilisation of existing test equipment and support systems can reduce substantially the life cycle cost and enhance product system availability. Training costs are reduced and the impact of the new system in production and use is minimised.

The long term benefits of introducing new standards should also be considered. Examples include standardisation of programming languages, the use of standard multi-test equipment, standard test specification languages such as ATLAS and standard test connection interfaces. Standards are also evolving in the area of design for testability, notably as applied to digital electronic circuits and systems. The Computer Society of the IEEE has set up a Working Group to define a Testability Bus Standard for electronic devices, printed circuit boards and systems. The bus proposed provides an analogue and digital input/output interface that can be used, in conjunction with a circuit's normal primary input/output pins, to enhance testability via circuit partitioning and improve controllability and visibility (see also 5.6.2).

Attention is also drawn to the IEE publication "Guidelines for the Documentation of Software in Industrial Computer Systems" which provides a valuable basis for the preparation of in-house standards on many aspects of project documentation.

2.5 Data Collection, Analysis and Documentation

A testability programme cannot be totally effective unless a documentation system is in place for the **systematic tracking and evaluation of testability effectiveness beyond the development phase.**

The objective is to plan for the evaluation of equipment testability in operational and maintenance environments.

Much of the collection and analysis of data and resulting corrective actions may occur after the end of an equipment supply contract and may be performed by personnel outside that contract. Nevertheless, **it is essential that the planning for this should be initiated during the Full Scale Development Phase and included in the overall critical design review.**

Separate plans should be prepared for testability data collection and analysis during

the production phase

the deployment phase

The plans should clearly define which data are to be reported in various documents, such as test effectiveness reports, production test reports, factory acceptance test reports and field support reports.

2.6 Partitioning and Sub-system Control

An important consideration in the design of any product is the partitioning of the whole into sub-units which can be separately manufactured and tested. Partitioning should take into account the manufacturing, testing and maintenance strategies.

One objective is to enable testing in manufacturing to be carried out as simply as possible and at an early stage in the assembly process. Another is to ensure that field maintenance is simplified by an optimum choice of the size and function of replaceable parts. Further advice on the partitioning of electronic assemblies is to be found in Section 5.5.1 and Section Seven contains a number of relevant rules.

2.7 Condition Monitoring

The ability to predict a failure or the need for inspection in a system or equipment during service has obvious benefits in cost reduction and safe working. The subject is dealt with in some detail in Section 6.5 under Health and Usage Monitoring. As hardware for sensors and data analysis decreases in cost, the opportunities for making improvements in product performance and customer acceptance should not be underestimated. The techniques are particularly advantageous for products incorporating moving parts, or working in a hostile environment. Consideration of the advantages of introducing condition monitoring should be included in the testability assurance programme.

2.8 Safety Considerations

This guide refers to the recommendation of suitable tests and test procedures and in no way absolves the supplier of equipment or the user from statutory obligations relating to health and safety at any stage of the manufacture or use. Provision should be made for the safety of individuals engaged on testing and care taken to prevent unnecessary exposure to hazardous test situations. Particular emphasis should be placed on all aspects of safety and every effort should be made to anticipate hazards caused by equipment malfunctions.

When considering safety matters, designers should have regard to the level of skill and experience of the test and maintenance staff likely to be used for such testing and the nature of the hazards involved. Safe operating procedures must be maintained at all times during testing.

Particular care needs to be taken in testing complex systems which combine the use of a number of engineering disciplines. For example, an electronics circuit engineer may not appreciate all the hazards arising from the use of high voltages, high energy mechanical forces, pneumatics or hydraulics.

The reader's attention is drawn to the technical guidelines contained in "Programmable electronic systems in safety-related applications", published by the UK Health and Safety Executive.

2.9 Test Economics

The starting point for an economic analysis of test is knowledge of the Expected Fault Spectrum (EFS), target yield and volume flow (throughput). The EFS details the types and distribution of anticipated failure mechanisms. There is little point in targetting a test program on a range of failure mechanisms with low probability of occurrence if those with higher probabilities remain untested. The EFS is therefore essential knowledge and yet, all too often, is either missing or only known about in a vague manner. This technique is sometimes known as Failure Mode and Effects Analysis (FMEA).

The term "yield" is defined to be that percentage of a collection of items that are considered to be without fault. Mathematically, yield is a probabilistic quantity and can be derived from a knowledge of the EFS. Automatic Test Equipment (ATE), where employed, can be used to monitor actual yields for comparison against target yields but the figures obtained should be tempered by the known ability of the test program to detect all failure mechanisms.

Using EFS and throughput data, an economic analysis can be carried out to determine the total cost of testing a batch of Units Under Test (UUTs), and hence the cost of test per unit. This figure provides a basis for judging the cost-effectiveness of a Design-For-Testability change.

In summary, the major cost factors to consider when carrying out a cost analysis are:

Cost of preparing a test specification

Cost of preparing a test program with known fault coverage and diagnostic performance, plus the associated software tools and hardware environments

Added design cost and manufacturing unit cost of providing testability features

Cost of applying the tests on test equipment and ATE, including fixturing and capital costs

Cost of carrying out repair activities

Cost of re-testing repaired unit

Cost of recruiting, training and maintaining the various personnel associated with all these tasks.

Perhaps the most important factor to be considered is the cost of not carrying out tests and the consequent implications on project delays and product deliveries.

The ultimate target of a DFT change is to drive down the sum of these costs as far as possible without compromising the ability to measure yield properly and to ship quality products.

In considering testing costs, the cost of false calls or no-fault-found conditions should not be overlooked. A false call occurs when a component or part is reported as faulty without a defect being found.

2.10 Training

The techniques and skills required by designers and test engineers to meet testability objectives are not generally covered by academic courses. Management must therefore provide adequate opportunities for training in

Management of testability

Design-for-test techniques

The use of test generation and assessment tools

The application of manual and automatic test equipments for production testing and in-service maintenance.

Attention is also drawn to the annual Vacation Courses organised by the Institution of Electrical Engineers and by various academic and commercial organisations. These provide a valuable training in designing for testability.

Another valuable training aid at a general level is the video film "Testability", produced by the Electronic Engineering Association for the Department of Trade and Industry and available from the latter at Room 2895, Millbank Tower, London SW1P 4QU. The current price is £23, including VAT. This film shows products being designed for testability and being tested in production and in service. It explains the importance of designing for testability and getting the message across to managers and engineers in industry.

2.11 Use of Expert Systems

The application of expert systems or Intelligent Knowledge-Based Systems (IKBS), as they are sometimes known, to the solution of test and fault diagnosis problems is becoming of increasing importance with the growth of experience in the development and use of such systems.

The simpler expert systems consist of a structured knowledge base of rules derived from human experience in carrying out a similar function. For example, an expert system developed in the USA for supporting the diagnosis and repair of a diesel electric locomotive contained 530 rules, chiefly representing the accumulated knowledge of a senior field service engineer. The coverage of such systems is limited, however, to failure modes already experienced.

More sophisticated expert systems can be derived from a detailed survey of the logic and operation of the design in which all possible failure modes are analysed and incorporated. Such an analysis can rightly be regarded as part of the designer's testability responsibility; it enables design for testability and design for maintainability features to be incorporated in the design in order to improve the performance of the product.

Expert systems have obvious advantages in reducing the reliance on the training and maintaining of human resources. This is particularly important as the reliability of systems increases and the opportunities for learning and developing maintenance skills on the job decrease.

The problems arising in testing expert systems are complex and have not yet been solved.

Section Three: Assuring Testability

3.1 Introduction

The testability of a component, unit or system can only be achieved if the objective is first defined and then steps are taken during the design phase to make adequate provision for measurement.

This Section covers the following topics:

> 3.2 Definition of testability
>
> 3.3 Rules for assuring testability
>
> 3.4 Assessment of testability

3.2 Definition of Testability

The concept and definition of product testability involves considerations of testable entities and transfer functions which have to be defined before Testability can be defined:

A **Testable Entity** is an entity whose transfer functions can be defined and verified via test points which are accessible to the test facilities provided either internally or externally.

A **Transfer Function** is the relationship between an output response of a testable entity and any input stimulus or set of stimuli that produced the response.

Testability is the inherent property of a Testable Entity which facilitates the verification of its defined properties and performance. This leads to the diagnosis of malfunctions in terms of its characteristic transfer functions, in a timely and affordable manner.

The following corollaries can be derived:

> An Untestable Entity has transfer functions that either cannot be defined or cannot be verified due to lack of access to test points, or lack of suitable equipment.
>
> To achieve a high level of testability for a product as a whole, each of its replaceable component parts should be Testable Entities.

Initialisation, Controllability and Observability

To achieve product testability, provision must be made in a design for the presetting and initialising (initialisation) and the provision of stimulating inputs to the unit under test (controllability) and for the verification of response outputs (observability).

> **Initialisation** is the setting of an entity into a known state.
>
> **Controllability** is the facility whereby an entity can be set into a defined state by the injection of test inputs at some convenient input or node.
>
> **Observability** is the facility whereby the overall and internal transfer functions of an entity can be verified.

3.3 Rules for Assuring Testability

In order to provide high levels of testability, the concepts and details of product designs must fulfil a set of testability rules. The rules can be divided into two types, General and Technology-Specific.

General Rules relate to fundamental concepts of designing for testability and apply to all product designs irrespective of the particular disciplines that are used to design, manufacture and test a product. (Such particular disciplines can relate to the mechanical, electrical, electronic, hydraulic, pneumatic, optical or other characteristics of a product).

Technology-Specific Rules arise from particular design, manufacturing and testing disciplines. They may relate to:

The nature of the product

The environment in which a product has to function

Limitations imposed on production costs

Limitations imposed by the available test equipment and skills for both manufacture and the in-service maintenance of a product.

Checklists of General and Technology-Specific Rules can be found in Section Seven of the Guidelines.

3.4 Assessment of Testability

3.4.1 The use of checklists

The required level of product testability needs to be defined as part of the product specification and its achievement assessed during the design phase. For digital electronic circuits, testability analysis tools are available which enable the ease of creating an effective test program to be predicted by an analysis of the circuit structures (see Section Five). In general, however, subjective methods have to be used. One such method is to assess a product design against a set of design rules (checklists) which are weighted and awarded a number of points according to their effectiveness and applicability to the product and its intended use. Each weighted rule then contributes a score to the total figure of merit for the design. A simpler and more subjective method is to use such a checklist of appropriate rules but without applying a weighting factor.

The method of developing and using a weighted checklist approach is explained in more detail below.

Checklist preparation

A checklist of testability-related design issues of relevance to the particular project is prepared and a method of quantitively assessing the degree of achievement for each entry is proposed. This scoring method is then negotiated for agreement with a reviewing authority, and should be finalised prior to the preliminary design review.

The basis for a suitable checklist can be prepared by using the lists of testability rules which are contained in Section Seven of these Guidelines, although these rules should not be regarded as exhaustive. other rules may be derived from an analysis of problems encountered on previous projects.

Checklist scoring

As the design progresses, each checklist entry is examined and the design scored for testability compliance. The weighted and summed scores give a figure of merit for the design. The testability features of the design are developed until an acceptable predetermined threshold value is reached.

Threshold value determination

The project manager or requiring authority should assign an acceptable threshold value for the testability figure of merit. More important than the absolute value is the negotiation of the weighting factors which modify the individual checklist entry scores according to their importance.

It will be appreciated that the checklist scoring method is, in part, subjective and can only assess relative testability. However, it is simple to apply, and with consistent application over a number of products, can prove a useful input to design reviews and audits.

Further guidance on the generation and application of testability checklists and weighting factors may be found in Appendices A and B of MIL-STD-2165. In addition, IEEE Standard 982 "Software Reliability Measurement" gives guidance on the assessment of test coverage for software.

This activity is essentially an iterative process throughout the design, development and early production phases.

3.4.2 Analysis of test effectiveness

When test programs have been generated, their effectiveness in detecting and locating faults needs to be assessed. The fault coverage of test sequences is not easily measured directly due to the difficulty and expense of introducing, in turn, a very large number of hardware fault conditions. Software simulation can be a useful tool as it enables a large number of digital and logic faults to be simulated and the effectiveness of test programs in diagnosing these faults can then be checked. The test programs can be further developed or the design modified until the required level of test performance is achieved.

Failure mode and effects analysis (FMEA) is an effective if somewhat laborious procedure which is applicable to complex engineering systems. In this procedure, all possible failure modes that can be foreseen within specified ground rules are documented, using cause-consequence and fault tree analysis techniques. Each failure mode may then be assessed according to its assessed criticality category and probability of occurrence. The total process, known as failure mode, effects and criticality analysis (FMECA) can then be used as a basis for checking the effective cover of the test and maintenance schedules as well as in system reliability prediction.

3.4.3 Demonstrations of test quality

A demonstration of the effectiveness of the fault detection and isolation capability should be required as part of the system design review. This should cover both the higher level off-line and sub-unit testing and the on-line site maintenance level procedures.

Factors to be assessed are

Effectiveness in detecting and isolating faults

Time taken to detect and isolate faults

Time to effect repairs and retest

Test equipment interfaces and fixture requirements

Test software issue levels and documentation.

TECHNICAL CONSIDERATIONS

SECTION FOUR : Software Testing and Testability

SECTION FIVE : Electronic Devices and Assemblies

SECTION SIX : Electrical and Electromechanical Products and Complex Systems

Section Four: Software Testing and Testability

4.1 Introduction

The testability of software depends on its structure, on supporting documentation and on a suitable environment for testing. A testable unit can be a procedure of any complexity, or a larger unit such as a task or complete program. The testing of software units depends on their having:

> defined transfer functions (see Section 3.2) expressed in terms of their inputs and outputs (given in the design documentation, in a suitable language or notation);

> a testing environment in which the unit under test (UUT) can be examined or executed in a controlled way.

In addition to considering the basic principles for designing software, this Section also covers the techniques for the testing of software in some detail, since it is important for readers to understand the basic differences between the testing of hardware and software. The Section addresses not only the testing of software, but also the testing by software of its environment (see 4.5.3). Whatever test methods are used, the importance of ensuring at each stage that the specification is unambiguous, accurate, complete and with a clearly defined test objective cannot be overstressed.

The following topics are considered:

> 4.2 General considerations
>
> 4.3 Stages of software testing
>
> 4.4 Test techniques
>
> 4.5 Designing for testability
>
> 4.6 Test planning and project management
>
> 4.7 High integrity software
>
> 4.8 Test tools

4.2 General considerations

Software is the embodiment of a design specification for execution by a processor and so might not be expected to fail or decay in performance characteristics during service. Ideally, the software that exists at the start of the operational life of a system should be that which exists at the end, assuming that no modifications have been implemented in the interim. However, users of software well know that software can "fail" and deteriorate in performance during service. This can be due to:

> direct design faults
>
> consequential design faults
>
> inadequate protection against external faults.

Therefore, there are various aspects of software that require careful consideration for its effective design, production, testing and operation.

One notable feature of software is its flexibility which includes the potential for faults to be introduced at each stage. In order to retain the benefits of a well specified, designed, produced and tested entity it is important that the software is supported by adequate documentation for the user, maintainer and developer. This should be backed up by well defined and managed procedural arrangements between the software supplier and user as well as within the user's own establishment.

It should be noted that software differs fundamentally from hardware as far as testing is concerned.

Hardware must undergo two distinct phases of testing, both of which are significant:

> Design validation—where the functional characteristics of the product and design specification are checked for correctness and completeness against the original requirements.

> Product Testing—where each manufactured product is checked for correctness and completeness against the product design specification. These tests (or a subset of them) are conducted at various stages during the product's life cycle (manufacture, commissioning, operation).

For software, product testing is trivial since each final deliverable product is a precise copy of the original (achievable by the application of configuration management procedures). Only the design validation phase is significant for testing the product.

Testing can never prove that software is correct. Instead, it gives us confidence in the software tested by a double negative process: we try diligently to find faults, and (hopefully) fail to discover any. The confidence is only justified if the search is thorough. A fault that causes problems five days or five years later was there all the time, but our tests were not thorough enough to find it. Software testing is never conclusive; that is, it is never known how many errors in the software remain undetected.

Section *4.3* provides guidance on the design validation and testing of software and is summarised below. However, the reader's attention is drawn to the "Guidelines for the Documentation of Software in Industrial Computer Systems" published by the IEE which covers this subject and many other aspects of software procurement and documentation in greater detail than is possible in this document. It is recommended reading on this topic.

Starting with the user's specification of requirements and progressing through the various stages of production, the software design must be checked carefully and regularly for correctness at each stage. The ease with which these checks may be carried out will be greatly influenced by the design methodology employed and the resulting software structure.

The later stages of testing require the software to be loaded into memory and executed by the processor. These stages can be complicated by the fact that the hardware and the software which is not under test (e.g. operating system, standard library modules, etc.) can influence the results.

Initial software testing is confined to the computer test environment, with software modules simulating external stimuli and responses. Integration of the software with the system hardware is achieved by a controlled progression towards the operational environment.

The next most important stage is to ensure that the operational software remains uncorrupted before, during and after execution and that it is a precise copy of that which was tested and deemed to be correct at the outset. This can be achieved by building in certain safeguards in the software design together with good management and housekeeping of the distribution and use of the software, using established configuration control procedures.

4.3 Stages of software testing

Software which is error free, that meets users' requirements, that is tolerant of potential hardware failures and which is capable of future development with minimum risk and effort must evolve through a series of stages, each of which is dependent on the previous stage being thoroughly and accurately completed. **Therefore each stage of the software design must be fully documented and tested against the requirements of the previous stage.** Each level of design must address the functional requirements together with the means of design validation and "in service" testing requirements.

The generally accepted method of software design is the "top down" approach; starting with the overall objectives, the design process defines functional modules which, combined together, meet these objectives. The modules will equally be capable of being tested in isolation and the process of sub-dividing into smaller, more detailed functional modules continues until sufficient design detail allows coding to commence.

The normal phases or levels involved in software production are listed below but the reader should note that a degree of flexibility is permitted to suit the complexity of a particular application.

Feasibility

Requirements specifications

Functional/facilities specification

System specification

Software system specification

Sub-system design

Module design

Unit Test

Integrated test

Sub-system test

Acceptance test

User test

Cutover to operational usage

Testing/assurance of correctness (i.e. exact copy of tested code and/or data)

Only the final design phase results in software that is loadable and testable as a software entity. Consequently, it is important that documentation of suitable quality for validation testing is produced at each intermediate stage.

The following describes each design stage, together with the design considerations that should be applied at that stage in order to make validation possible.

Feasibility Study

This stage involves the identification of the problem(s) and considers in general terms the suitability of a computer (system) solution. Alternative solutions are also considered and projected costs and benefits are established. These studies (sometimes supported by research and development projects and other trials) allow the most effective solution to be identified.

The result of the feasibility study should be the identification of the technology and methods to be employed and the qualification of the achievable objectives to meet the requirements of the manufacturing and maintenance strategies.

Tested by referral to project sponsors, etc.

Requirements Specifications

At this stage the user's requirements of the total system (and thus the software) are itemised and defined in absolute terms. This should be a detailed, unambiguous statement of what the system is expected to do and the type of tests to be carried out to check that the system meets its requirements.

Tested by reference to conclusions of Feasibility Study and Project Sponsors.

Functional Specification

The Functional Specification is usually prepared by the supplier and agreed jointly by the supplier and user. In addition to fully describing how the system will meet the user's requirements as set out in the User Requirements Specification, the Functional Specification will define attributes of the system concerning testability, reliability, maintainability, adaptability, security and operational considerations.

Since the final agreed version of this document usually forms the basis of the contract between user and supplier, **it is of vital importance that the procedures for testing the software at all stages of its development, system integration and operation are stated clearly at this stage.**

The functional attributes are checked for correctness and completeness against the User Requirements Specification and by reference to the Project Sponsors.

System Specification

The System Specification decomposes the Functional Specification into Hardware and Software System Specifications and also specifies the integration into the operational system.

Software System Specification

The Software System Specification consists of a documentation set which describes the structure, functions and content of the software system and the plan for its progressive integration into the operational system. This specification is normally produced by the supplier. Depending on the level of expertise available to him, the user may (or may not) choose to monitor and approve the content of the Software System Specification. Whichever is the case, **the user or sponsor should make sure that appropriate techniques and procedures are implemented to allow checks and controls to be applied.** He should also establish procedures which will enable changes to the contents of this document—or lower levels of documentation—to be referred back up the documentation hierarchy for validation and (if necessary) modification.

Tested by reference to Functional Specification.

Sub-System Design

For smaller applications, this stage can sometimes be omitted. For large systems, it is often desirable to divide the system into a number of sections or sub-systems. A sub-system typically will be capable of independent development, implementation and testing and integration with its operational hardware. The relationships between sub-systems will be formally specified and considered inviolate.

The criteria used in deciding the boundaries of a sub-system can vary widely. They usually are functional but additionally can relate to method of implementation (e.g. implementation as a series of phases) or software generation methods (e.g. multiple software development teams). Special care is needed in defining the interfaces between sub-systems.

Tested by reference to System Specification.

Module Design

This stage relates to the design and coding of the individual software modules. The designer/programmer should give thought at this stage as to how the module may be checked for correctness against the System Specification and then incorporate any necessary features into the design. The design should also incorporate any requirements laid down in the System Specification for integrated testing.

This stage of a software project is usually the most costly It is of vital importance, therefore, that all ratification of the System Specification has been completed before commencing this stage.

Tested by reference to System (or Sub-System) Specification, and extensive static testing by the compiler if a suitable programming language is used.

Unit Testing

This stage represents the most detailed level of testing to which a software system is subjected. Various methods of testing and design validation may be employed at this stage to show that an individual module is correct and complete against the structure, function and content stated in the System Specification (see Section 4.5 for details).

Tested by reference to Module Design document and System Specification.

Integration Testing

Integration Testing is the next stage in the test sequence to which the final software is subjected. Two or more modules that have successfully completed the unit test stage, and which are mutually related, will be combined and subjected to appropriate tests.

Failure to pass these tests will necessitate correction and re-test at module level and a repeat of the integrated tests.

Tested by reference to System Specification.

Sub-System Test

One or more sets of modules that have passed the integrated testing stage are combined to form the subject of a sub-system test. The System Specification will have specified the tests to which the sub-system must be subjected and the method of its integration into the operational system. Failures at this level of testing may require correction at module level and re-validation through the test hierarchy back to this stage. If, however, the tests uncover an error or inconsistency in the design as specified in the System Specification, then corrections at this level must first be referred up through the documentation hierarchy for ratification **before** being propagated to the module design level. Correction and re-test of the modules as previously described is, of course, obligatory.

Tested by reference to System Specification and, if necessary, the Functional Specification.

Acceptance Test

This represents a major contractual milestone of software procurement, and is normally the first stage of testing with which the user is directly involved. The tests are designed to demonstrate the software system's compliance with the agreed Functional Specification. The entire software in its final operational environment is tested at this stage.

Tested against Functional Specification.

User Test

User testing is a period of user familiarisation rather than exhaustive testing. It is sometimes alternatively known as cold commissioning. The user will have "hands-on" experience of the system and will assess the system's performance against a subjective set of criteria (perhaps the Requirements Specification). This is a period of user confidence-building rather than testing.

Very often operator training is incorporated in this stage.

Tested against User Requirements Specification

Cutover to Operational Usage

The initial period of live operation of the software is sometimes referred to as "Hot Commissioning". The system's performance is closely evaluated under all operational conditions and any tuning facilities adjusted for optimum performance.

Testing/Assurance of Correctness

Having completed the foregoing stages of testing, the final software product should be clearly identifiable and the means of confirming that any copies of this product are precise should be implemented. Distribution of copies of the software should be carefully controlled such that retro-fitted modifications or enhancements can be applied to all copies. Version and date codes are strongly recommended for this purpose and the means of establishing the identity, version and date of an installed software product should be provided.

4.4 Test techniques

When formulating plans for carrying out the test task, one should consider the various types of testing that would be most appropriate to the application, taking into account the effectiveness of each type, the likelihood of occurrence of the faults it detects, and the risk of faults being present that the test does not detect. The choice may be constrained by the availability of hardware. or software facilities. However, this section describes the most common types so that each may be considered when examining what might be tested. A problem associated with the selection of appropriate testing techniques is that there are no means of quantitatively assessing the effectiveness of a testing technique or tool. Practical experience in the use of various techniques provides the best guidance. Some qualitative assessment methods are discussed in Section *4.8.4.*

Testing can be considered from two major viewpoints—Static Testing (or Checking) and Dynamic Testing.

Static Testing

In static testing products are compared with each other to verify that distortions have not been introduced during the design process, e.g. where details of the system specification documents are compared with the requirements specification to verify that each item accurately translates information from the previous document. Static testing is performed without any execution of test data. The requirements, design documents and code are analysed, either manually or automatically, without actually executing the code. All design reviews, inspections and desk checking of code are forms of static testing. Methods of static testing are described further in Section *4.4.2.*

Dynamic Testing

The use of test data as input to the programs may emphasise different aspects of the design process. If the emphasis is on meeting only the "requirements" for the design step then it is called Black Box testing. In this case, the tester is completely unconcerned about the internal behaviour and structure of the program, and is only interested in finding circumstances in which the program does not behave according to its specifications. Usually such test data is defined solely from the specifications. If the emphasis is on the internal structure of the program only then it is called White Box testing, in which the tester derives test data from an examination of the program logic and ensures that all program paths and limits are exercised.

The fundamental difference is that **static testing** refers to the program text (assuming correct implementations by compiler, utilities and target computer), whereas **dynamic testing** refers to the executing form of the program in the target computer. Associated with this difference is a contrast between the kinds of expectation for the two methods. In static tests, the expected outcome concerns structure, relationships and behaviour; whereas in dynamic tests the expected outcome concerns particular values of output variables. These differences are important because they determine what must be stated on the testing plan documents as the expected outcome from each test.

4.4.1 Black Box and White Box Testing

As indicated above, test methods may also be classed as either Black Box or White Box.

Black Box testing

Black box testing is simplest. This treats the software under test as completely invisible, defined only by its specification. The test cases and the number of cases to be tried are derived entirely from the specification, by considering the structure of the specification as the criterion for partitioning the data space. Black box testing gives confidence not only in the software but in the specifications from which it was derived: the above analysis (which is quite different from that carried out by the designer) exposes any weaknesses or ambiguities in the specification itself. This should normally be the minimum level of testing applied to any software, but is rarely sufficient.

White Box testing

White box testing, or program structure testing, is more thorough, the test data being derived from the program's logic. It takes account of the structure of the program under test, typically thinking of it as a directed graph or flow-chart where nodes correspond to individual actions, and arcs correspond to the order in which they are performed. A path is a contiguous sequence of arcs and its length is the number of arcs in the sequence. The flow graph may contain loops, so there may be paths of infinite length. There are several degrees of white box testing, depending on the lengths of paths in the flow graph. The coverage of a test set may be expressed as a Test Exploration Ratio, being the proportion of paths of a particular length that are actually covered in the tests. **Node-testing** means testing all nodes in the graph, in other words all individual actions (basic blocks, sequences of non-control statements). **Arc-testing** means testing all paths of length one, in other words all sequences from one node to another. It is distinctly more powerful than node-testing.

The NATLAS standard and method for Software Unit Testing is a form of white box testing that gives particular attention to the conditions in the nodes at forks in the flow graph, and at the extremities of data regions.

White box and black box testing may be carried out simultaneously by the same tester, but generally extensive white box testing is carried out before black box testing is attempted; this is especially the case with large applications.

4.4.2 Static testing

Given the program code, there are a number of tests that can be carried out before the code is run with test data. These tests are known as static tests. They take the source code as input, and produce a number of analyses of the code which can be very useful in assessing whether the code will carry out the functions and operations intended. Of course, such comparisons depend on the existence of separate formulations of the intended effects of each part of the program.

One form of analysis is control flow analysis. The various paths through the code are extracted as a directed graph, and from this any unexpected control flow, such as unreachable code, jumps into the middle of loops, and so on, are revealed. (Most modern programming languages seek to avoid such problems in the source).

A second form of analysis looks at data use, to identify anomalies such as the use of a variable's value before it has been assigned, the re-assignment of a value before the previous has been used, and so on. A more sophisticated form of this will analyse information flow, to show which input values are used in computing the values of particular output values.

A third form of analysis attempts some form of proof of correctness. A specification of the required function in some formal mathematical notation is required, and from this the compliance of the code with the specification is assessed, assuming that the program is implemented correctly by its compiler, utilities and target computer. This kind of analysis is limited to the size of program unit that can be handled, but is important in safety critical applications.

4.4.2 Dynamic testing

Dynamic Testing, both black-box and white-box, can be carried out in a variety of forms. All methods assume that the test cases and program code have been constructed; the test comprises running test data against the code produced, and comparing the actual behaviour with the expected behaviour (see Section *4.8.3*).

(a) **Top-down Testing**

This approach is primarily concerned with test execution and nominally begins at the top or the control level of the software hierarchy. The units which comprise the control path, i.e. the links between the control level and the processes at the first level down, are integrated and this serves as the driver for the next level. The second level takes units through level 2 and integrates them with the control level above. Data is passed directly from the actual units above and units at the level below are simulated with "stubs" which simulate performance of the lower level software units. As software becomes available at the third level, it replaces the stub for the software using data passed from above until all levels of the hierarchy have been completed.

(b) **Bottom-up testing**

Under this test approach a driver which simulates the execution environment of the unit is written and used to exercise the software before it is included with other elements of a sub-system. The drivers are the key to the success of bottom-up testing. They are often difficult to write, must exactly emulate the execution environment and are then discarded when the unit passes up to the next level. This is a technique which may be used to integrate tested units into an executable configuration.

(c) **Combined top-down and bottom-up testing**

In reality a combined top-down, bottom-up approach at all levels of testing is most frequently applied. A software hierarchy is defined as if the software were to be integrated in a totally top-down fashion. Before software is integrated into the hierarchy, it is exercised using a driver which simulates data and control inputs. When all interfacing units have been exercised with drivers the unit is incorporated into the hierarchy and the software is driven by the driver at the next level up in the hierarchy.

(d) **All-path testing**

This method ensures that each and every basic path through a program unit is tested by its own item of test data. This is not exhaustive testing but it allows a carefully pre-defined level of testing to be achieved without using an excessive number of test cases.

Combinations of paths are exercised during the test process, although not every possihble combination can be covered. By ensuring that every path is exercised, a high proportion of errors can be eliminated. Data path diagrams can be drawn as a test design aid in order to enumerate the test cases that must be provided.

(c) **Thread testing**

This is a technique of integration that can monitor the integrity of interfaced software units as they respond to an external stimulus. A thread is a set of program units which, when executed, accomplishes a specific set of functions and results in a single or related set of outputs. Thread tests are applied when the internal integrity of the software has been established in unit testing. Testing begins with the exercise of single threads and then progresses to the testing of multiple threads.

(f)   **Requirements testing**

The technique consists of building a model or prototype system, not with the intent of using the system in operation, but to test and confirm the understanding of the true requirements. This is useful when requirements are little understood and some experience of a working mode is necessary to establish true requirements, particularly in relation to the design testing of user interfaces.

A similar approach can be undertaken to test out critical aspects of design. The basic technique consists of building a simplified model of a selected design property and using the model to test the underlying design. Such models are used to test database design configurations, transaction sequences, process sequences, response times and user interfaces.

4.5 Designing for Testability

In designing testable software, the same basic considerations apply as for the design of testable hardware and systems (see Section *1.4*). Some particular problems related to High Integrity Software are dealt with in Section *4.7*.

4.5.1 Sequential programs

Designing testable software involves the employment of rules of good practice, aided where appropriate by the application of design tools.

Section *7.1* gives a list of those design rules which are equally applicable to software, hardware and systems. More detailed rules, specifically relevant to the design of software may be found in Section *7.3*.

4.5.2 Concurrent programs

For more complex systems containing concurrent programs in which several tasks interact, correct structuring is needed to avoid serious testing problems. One strategy is to separate out the distinct sequential processes, and encapsulate the inter-task communication facilities of the system (i.e. the data interchange) as separate entities. This allows the information transfer to be presented to the individual processors in an orderly sequence. In the MASCOT design method, for example, activities and data exchange entities are distinct. Information is added and extracted from such entities by using access procedures. These access procedures perform non-interruptable changes to data within the entity.

This type of partitioning permits individual testing of:

the inter-task communication facilities (with simple activities)

the sequential processes (by executing them individually, using test data at the communication interfaces)

groups of processes (by using test data at the boundaries of the group).

Furthermore, it is relatively easy to add on-line monitoring into such a system by inserting spy-points into the inter-task communication facilities.

4.5.3 Diagnostic facilities and built-in-testing

It is advisable to incorporate adequate levels of diagnostic software to support checking and testing of the hardware and software during development and commissioning. These should remain within the operational system to facilitate the routine maintenance and the diagnosis of unexpected failures. Care should be taken to ensure that these features cannot affect normal operation (e.g. there could be a separate diagnostic mode). The diagnosis facilities could include:

Information on the software state prior to a 'crash' such as:

the register and stack contents

the currently active program (in a multi-tasking system)

the failure point in the program

a trace of the call sequence prior to the crash

Facilities to check the installed software:

software version numbers

program contents checks (e.g. sumchecks)

Checks on the computer hardware:

instruction set tests

RAM memory tests

manual read/write to memory

Checkout facilities for the external interfaces to support fault tracing and commissioning:

read values from the interfaces

set values on the interfaces

Software, as well as satisfying the application requirements placed on it, can equally well be designed to include a significant proportion of BIT in the integrated system of which it forms a component part.

In any computer system a large degree of BIT will be embedded in the hardware and operating system. The application software design should complement and support this. Ideally, the BIT of any system hardware and software components should be considered and specified as mutual requirements on both. The decision on how best to implement these may then be made with regard to cost, complexity, reliability, maintainability and any other criteria considered relevant.

The overall objectives of BIT should be established first. The primary objective of BIT should, of course, be to establish a go/no-go, healthy/faulty status of the system or sub-system. BIT may also be required to identify the actual fault and to pin-point the offending hardware component or software module or to indicate deterioration in performance. The required precision of such a diagnostic feature is an important design decision since the resulting reduction in system downtime and maintenance cost may have to be achieved at the expense of increased capital cost and reduced operational speed and efficiency.

The maxim "the design of the software system should be suspicious of its environment" is an excellent philosophy for all aspects of software design. For the software to be "suspicious of its environment" it should be designed to consider the following possibilities:

(a)   Corruption of program instructions or data by failure of the memory containing it (whether it be volatile or non-volatile);

(b)   Corruption of the program instructions or data as a result of erroneous action by itself or other software;

(c)   Presentation of data such that the software is caused to operate outside its design limits. The data may cause this by being out of range or by being logically incorrect in combination;

(d)   Corruption of program instructions and/or data during transmission to/from other memory devices. Movements within its own parochial memory are also potentially prone to failure.

Item (a) is usually catered for by the processor hardware in combination with the operating system or supervisory software. Memory parity checking will detect single bit errors and by employing more complex algorithmic methods, multiple bit errors may also be detected and may even in some cases be automatically corrected by the checking circuits.

Item (b) is not detectable by the parity hardware on the system memory since, as far as the memory read/write circuitry is concerned, a legitimate memory write cycle is being invoked. However it is possible to partition the memory and to detect and block attempts to address outside a program module's permitted area. This can be achieved by either 'hard' partitions with the detection logic entirely within the memory addressing hardware, or by a combination of hardware and (operating system) software to achieve a more flexible arrangement.

One additional method of safeguarding against either (a) or (b) is by employing a check procedure to ascertain the correctness of the stored program instructions and data before, after and if possible during execution. A checksum or algorithmically generated CRC (Cyclic Redundancy Check) is held with the instructions and/or data blocks and at appropriate times is recomputed and checked against the original. This procedure can impose severe system loading overheads, and therefore a tradeoff in requirements may be necessary.

The operating system usually provides for the handling of such errors and usually offers various options for annunciation and recovery. These will normally fall into the two categories of 'standard' or 'user-supplied'.

Item (c) is entirely the responsibility of the application software and therefore due regard should be made at the application software design stage to the effect and treatment of out-of-range data and invalid data combinations. Assistance in handling some situations may be provided by the operating system and the hardware. For example, arithmetic overflow can be trapped and passed to the application software, where the latter provides a program path for such eventualities which will be consistent with the system requirements. Another effect, often resulting from erroneous data, is "program looping" where the application software finds itself executing a program loop from which there is no logical exit. This condition can be detected by the operating system by employing an execution time limit for each module in the system and sometimes this is supplemented by employing a hardware "watch dog". A watch dog timer is a hardware device, running at the lowest processor priority level, which will provide a specified logic level (say high) if it does not receive an output from a servicing task within a specified time period. Thus, if any task is trapped in a loop, the watch dog timer service program is prevented from running and consequently (after the preset time has elapsed) the watch dog timer raises its output level. This in turn can interrupt the processor at a higher level or can initiate annunciation and/or safety circuits.

Item (d) is usually handled by the operating system in conjunction with the host computer, the peripheral controller (interface hardware) and the peripheral hardware itself (e.g. disc). Typically, this is achieved by means of parity checks on the individual characters of data and CRC checksums on data blocks before and during transmittal to the device together with a 'read after write' check to confirm that the data that has been written is and will be capable of being transmitted correctly back to the host. Depending on the sophistication of the CRC techniques employed a degree of error correction is possible. This will be handled by the hardware and the operating system, however the application software could make suitable provisions for such failures.

Erroneous reads and writes invoked by application software to other memory devices (due to a fault in the application program logic, say) will not be detectable by the above methods since such accesses will appear to be valid and normal so far as the standard checks are concerned. Consequently a further layer of protection is usually provided by operating systems such that the memory device is structured into discrete areas or files and access to them is normally only possible via the operating system. This allows a variety of protection features to be implemented to prevent unauthorised access by applications tasks. These features vary according to operating systems and device type but typically include such things as File Attach, Read Access only, Read/Write access, File Lock, etc.

It is important, therefore, that the application software design should utilise such protection features as are deemed necessary in order that maximum data security is achieved. Further it may prove necessary and desirable to build in checking features for specific files (or databases) that are the responsibility of the application software to administer such that the integrity of the data structure can be assured. For example periodic checks of chaining pointers in a Relational Database Structure could be carried out as a background process.

So far we have only considered the BIT of the application software and data. Of course BIT of the hardware components of the system and controlled plant could, and usually should, be part of the functional requirements of the system. Once again a degree of trade-off between the various counter-balancing criteria will be required. The following BIT will normally prove to be practicable:

(a)   Presence and operational status of computer peripherals such as discs, printers, digital and analogue I/O interfaces;

(b)   Presence and status of process interfaces such as transducers, actuators, etc.

Most standard computer peripherals will be catered for by the operating system which in turn can annunciate failures, construct statistics of failures, provide for automatic recovery from failures and also provide facilities to link to user supplied software for the application related consequences of failures to be handled.

Analogue and digital I/O, however, often requires special treatment. For example, the monitoring of an externally supplied voltage or current source which is maintained at a fixed standard level via the analogue scanner will confirm the operational status of the ADC and provide some degree of confidence in the analogue input system.

Analogue outputs can be checked by connecting an analogue output channel to an analogue input channel and arranging for the application software to periodically output a reference voltage or current and compare this with the value monitored on the input channels. Allowances must be made for allowable conversion accuracy errors of the DAC and ADC and also for the integration and settling times. The consistency and linearity of the analogue I/O system can be checked by outputting a number of discrete values throughout the working range. Digital outputs can be handled in a similar fashion by connecting outputs to inputs and performing readback checks.

4.6 Test Planning and Project Management

The Test Specification is vital to the success of software development.

Everyone should be aware of the need for **clear, complete, unambiguous and accurate specifications at the requirement, system design and program levels** (see IEE "Guidelines for the Documentation of Software in Industrial Computer Systems"). A carefully constructed Test Specification is essential. This will include the detailing of the testing plan, the testing requirements of the system and the allocation of responsibilities for meeting them. The production of a Test Specification so concentrates the mind that it often highlights omissions and ambiguities in the Functional Specification. Work on the Test Specification must therefore begin at the same time as the strategic system design and the contents must be reviewed during the detailed design and the programming stages to reflect all changes occurring during development. **Too often, the test specification is skimped** and plans, if any, are constructed too late just prior to the commencement of the testing stages. **This situation leads to inadequate testing and an uncontrolled, confused attempt at meeting project milestones.** Crisis management becomes the rule, totally supplanting the sense of order and visible project control. Testing becomes driven by desires not attributable to the quality of the system being developed but by timescale pressures imposed by disillusioned and often misinformed general management.

The contents of a Test Specification will vary according to the degree of complexity and size of a particular application. A typical requirement for a large industrial project would consist of the following headings:

> Scope and objectives
>
> Strategy and stages
>
> Test support—tools, facilities, hardware and software
>
> Test requirements/scenarios and cases
>
> Test procedures and practices
>
> Test control and review
>
> Test preparation—scripts and data
>
> Test execution, result checking and comparison
>
> Resource estimates and schedules
>
> Responsibilities for meeting testing requirements.

If enough thought is put into testability planning at an early stage, much of the heartache and frustration of testing can be avoided.

4.7 High Integrity Software

For systems required for "safety critical" situations, the Software Quality Plan will need to include special measures in order to achieve the required level of safety. Measures to be considered include the following:

> The use of a mathematical notation to express the Requirements Specification, subsequent levels of the Design Specification and to prove conformity between these levels
>
> The avoidance of common mode failures by the use of two or more independently designed and coded software programs to carry out the same functional task (fault tolerant software)
>
> Checking of the data output from all calculations or algorithms for correctness before further processing (defensive programming)
>
> Static analysis of the source and/or object code for completeness and correctness

Extensive testing for a large number of program cycles (e.g. 10^6 or more) using randomly generated input data.

There may be a requirement for such software to be assessed for certification purposes by an independent assessor.

4.8 Test Tools

Tools are vitally important for successful testing. Some of these will be of general utility throughout the development of software, while some will be specific to testing. In the subsections below the tools are described within broad categories (see also Appendix *B*).

4.8.1 General purpose tools

An important aspect of successful testing is the production of appropriate documentation:

> Test requirements and specifications
>
> Test plans
>
> Test procedures
>
> Test reports for particular test runs

as laid down in the quality manual for the software development project concerned.

The documents should be easy to produce. For this, document preparation tools should be available, to enable the easy production of the documents and their evolution through several rounds of change. This document preparation system should include limited graphics capabilities, so that forms can be produced, and flow diagrams drawn when appropriate. Associated with this document production facility should be the availability of skeleton documents for the particular classes of document concerned, with help text on-line in some form to prompt the document writer for the appropriate content.

Test documents and the other material produced as part of the test program are subject to revision, and the availability of a configuration management system is as important for test information as it is for other information produced on a project. Successive versions of documents, test-sets, etc. should be stored, and updates of these controlled to avoid duplicate or unauthorised changes. Also a change control system should be present.

An important aspect of testing is the tracing of requirements through to implementation, with a link into the test documentation and test cases so that for each requirement, it can be seen where it is satisfied in the implementation and how it is tested and when tested, with what outcome. The level of detail required here is much finer than is usually accommodated in configuration management systems, but with newer generations of tools, full configuration accounting to the level of individual requirements and tests would become possible.

4.8.2 Specification and design testing

During specification and design, various tools may help in validating the intended software prior to its implementation.

In order to establish the requirements and their embodiment in a specification, various prototyping tools might be used. These enable the customer or user to understand better their requirements and what is achievable in software. From these a specification would usually then be produced; this specification may itself be animated to produce a more sophisticated prototype. More abstract consequences of the specification may also be able to be deduced, possibly using theorem provers or proof assistants.

Once design has been commenced, the system may need to be modelled, perhaps to assess performance, or to look for design flaws such as deadlock using Petri-nets or similar.

4.8.3 Test execution

During testing, it is important to be able easily to repeat a test, and compare the results with those expected.

One facility for the easy repetition of tests is a test harness or test rig. This tool records the test inputs in a file, and enables test to be run by drawing their inputs from the file. The contents of the file could be produced using a special test-script language, or they could be captured during some interactive test of the software, with the facility to later switch to the file, and back again for more interactive input. The results of a test run would also be captured, and related in some way to the inputs for that run.

Associated with the use of a test harness may be the need for stubs to stand in for missing pieces of code. A suitable stub generator tool may be commercially available.

Having run a test, the results of that test need to be compared with the correct results. Obtaining the correct results can itself be tricky, and a frequent practice is to check out the results by hand for a first run, and thereafter compare one run with the previous. This requires a test comparator to check for differences. Frequently the file comparison facility, standard on most main-frame and mini operating systems, will suffice, but occasionally a richer comparison is required and a special tool will need to be written. Of course if differences are found, they could be due to an intended change of function, and will need hand checking.

In some instances, rather than designing individual test cases by hand, it might be easier to generate these. A special generation tool could be written, or a general purpose test case generator, for example based on finite state machines, may be available.

A special form of generator is an environmental simulator, which simulates the environment in which the software will finally run. Typically these would be specially constructed, but could be general purpose, being driven from some description of the environment.

4.8.4 Test quality assessment

It is important to assess the quality of a set of test cases. Quality is generally measured by some form of coverage, i.e. the proportion of the code that is exercised by the test set. The particular measure of code could vary from the number of statements, to branches, to distinct paths through the code. While the coverage of a test set can be assessed by hand, this is tedious. Test coverage tools "instrument" the code which is then recompiled and run to accumulate statistics during execution of the code using the test set. Afterwards the statistics produced are analysed to produce a report to show the coverage achieved. For statement coverage the tool is comparatively simple and there is no effective bound on the size of the software that can be assessed. At the other extreme, for path coverage, a lot of preliminary analysis is necessary to extract the paths, and a lot of information has to be accumulated during execution, so that there is an effective limit on the size of system that can be assessed in a single go.

A technique for coverage analysis, if the tools are available, is to use some form of bus-monitor. This is an extra piece of hardware which is attached to the processor bus and monitors the execution of the software without affecting its performance. Analysis tools are necessary to relate the addresses observed back to the actual code at its source level.

More elaborate assessment of a test set could use fault injection, to introduce faults deliberately into the software and then see whether the test set can find them. This method is very expensive because of the number of assessment runs involved, and thus would only be used in special cases. Where used, it could be appropriate to build a mutation tool to help in the assessment.

4.8.5 Performance measurement

Performance measurement requires additional special tools. It requires the ability to cycle through large volumes of test data, using test case generators, environment simulators, or specially adapted test harnesses. In addition it requires the ability to measure the performance of the code, not just overall, but also between internal points.

This could mean instrumenting the code as for coverage measurement, but this would affect the performance figures obtained. A bus-monitor would be preferable. Accurate timing, possibly over many repetitions to factor out noise, is necessary. Analysis tools common in experimental science would be necessary to make sense of the data, assess its significance, and display it using graphs.

Section Five: Electronic Devices and Assemblies

5.1 Introduction

This section outlines the current status of Design-For-Testability and test development techniques as applied to digital and analogue integrated-circuit devices and loaded printed circuit boards.

It is assumed that the testing can be implemented by using either Manual Test Equipment (MTE) or Automatic Test Equipment (ATE), according to the nature of the Unit Under Test (UUT).

The section is broken down into the following six sub-sections:

5.2 Objectives

5.3 Tools to support design

5.4 Tools to support testing

5.5 Designing for testability

5.6 Test equipment

5.7 Future trends

5.2 Objectives

5.2.1 Major test objectives

There are two major test objectives. The first is to apply a test pattern or waveform to a Unit Under Test (UUT) which is designed to detect the presence of any possible failure mechanism. This is termed the **detection objective.** The second is to be able to locate the cause of failure with sufficient accuracy so that an effective repair can be made, either to the UUT or to the process that caused the fault (or both). This is termed the **diagnostic objective.** Note that a test waveform may define a set of UUT input values (the stimulii) and the expected fault-free UUT output values (the responses). Generating test waveforms for either digital or analogue circuits is difficult in general and the tools and techniques for performance proving and fault analysis are more advanced for digital than for analogue circuits.

The detection objective is measured in terms of the ability of the test waveform to detect the presence of a range of fault-effect models such as the stuck-at-1 (s-a-1), stuck-at-0 (s-a-0) and stuck-adjacent (short-circuit) for digital circuits and out-of-limit response values for analogue circuits. Sometimes it proves difficult to correlate the fault effect with the failure mechanism.

For digital circuits, the diagnostic objective is limited by the phenomena of equivalent faults. These are faults which are collectively tested by the same tests and which are therefore indistinguishable in terms of identification as to unique cause of failure. The diagnostic ambiguity is further aggravated if the failure mechanism occurs along a member node of a closed feedback path. Under these conditions, all member nodes can exhibit faulty behaviour, thus making it difficult if not impossible to separate the cause from the effects.

The use of fault-effect models has been restricted so far to digital circuits. Modelling the effect of failure mechanisms on the behaviour of an analogue circuit is difficult and can only be achieved by using statistical techniques within analogue simulation software.

5.2.2 Other test objectives

Other test objectives are related more to the physical process of applying the tests to the UUT and are governed by:

UUT set-up and unload time

Throughput (number of UUTs tested/unit time)

Average fault-detection run time

Average fault-diagnosis run time

Average time to repair, etc.

Ultimately, the ability to achieve a set of objectives such as these is constrained by cost and time. Understanding the cost of developing, evaluating and applying a set of tests to a UUT is the key to deciding whether design-for-testability is necessary and, if so, the techniques to be used (see Section 5.5).

5.3 Tools to Support Design

5.3.1 Simulation

Tools supporting the design of complex circuits are many and varied, usually taking the form of simulation techniques with associated peripherals. Simulation tools are used to verify the correctness of the basic design in terms of functional behaviour, and to assess design parameters in terms of timing information where this is necessary.

For digital circuits, the assessment may indicate potential race hazards within a circuit.

Simulation tools are also used as an adjunct to the test generation process. When used as fault simulators, the tools can verify the coverage of a test vector, and discover additional faults that may be covered by a particular test.

Test vectors are bit patterns in digital words for digital circuits and constant or varying signal values for analogue circuits.

The range of simulation levels provided in any one simulator is usually restricted. The simulation levels that can be catered for are; circuit, logic structural, switch, functional, behavioural and fault.

Circuit level simulation represents perhaps the most accurate level that is achievable with standard circuit analysis methods. This type of simulation uses Kirchoff's voltage and current laws to deduce the circuit behaviour. Other simulation levels are greater abstractions from the reality with a consequent reduction in the accuracy of the result. Circuit level simulators can be used to determine the electrical characteristics of semiconductor circuits, including resistance, capacitance and inductance. Usually their use is limited to circuits containing up to 500 nodes and to determining critical design parameters, such as timing, which can then be abstracted and applied to a logic level simulation. They can require large amounts of CPU time and memory space. Circuit level simulation has yet to be fully developed for analogue circuits.

Logic level simulation is widely used to represent digital circuits by interconnected logic gates or Boolean functions. Logic simulators are computationally far more efficient than circuit simulators as the functional connectives are usually purely Boolean. Electrical signals are no longer continuously variable but discrete logic levels, and timing parameters such as delays, are represented by incremental values. Logic simulation can be used to verify the logical correctness of a design and to check for correct signal timing. They may be used for circuits containing many thousands of primitive gates, and are several orders of magnitude faster than circuit simulators.

Switch level simulation has been developed to overcome the fact that a logic gate is not necessarily a satisfactory representation of a circuit designed with MOS technology. MOS transistors are modelled as simple switches in on/off/indeterminate states for the bidirectional transmission of logic values. Ratio logic, signal strengths, and node capacitance are all catered for by this efficient form of simulation. Switch level simulators have the advantage over logic simulators of being derived directly from a circuit layout without the requirement for abstracting an intended logic function prior to verifying circuit behaviour. Many logic simulators now allow some switch level simulation capability.

Functional and behavioural level simulators provide the highest levels of simulation. They represent the algorithmic behaviour of the circuit or system. Pascal-like descriptions of the circuit can be partitioned into functional sub-blocks to describe the building blocks that define the system, such as Arithmetic Logic Units (ALUs) and registers. Being high level, these simulators are efficient computationally and may be used to define signal flow rates and detect potential bottlenecks. Again, some logic simulators allow for functional description of circuit blocks. Functional and behavioural level simulation is equally applicable to analogue circuits, but further development is needed.

It must be accepted that all simulations are abstractions from reality and therefore will not indicate the true behaviour of the circuit under all conditions. Many simulators will tend to be on the pessimistic side, for example, indicating potential timing problems where none exist. As with all software tools, the underlying principles must be understood to appreciate their limitations.

Fault simulators are used as an aid to test pattern generation for digital circuits to confirm or determine the fault coverage of a test vector. They are used to compare the logical response of a circuit containing a fault to that obtained when the circuit is operating fault-free. Given the general acceptance of the single stuck-at type of fault model (other fault types can usually be simulated in the later simulators), the fault simulator will provide a response for every test vector related to every individual stuck-at fault in the circuit. This becomes very inefficient for anything but small circuits, and many methods are used to reduce the CPU time. Parallel, concurrent and deductive techniques have been developed. Of these, the parallel and concurrent approaches are the more common.

Formal Verification methods are now in an advanced state of development although not yet widely available commercially. These tools are based upon formal mathematical analysis of the original circuit specification, and the physical circuit after design. Instead of performing extensive, sometimes exhaustive, simulations which are expensive to confirm the behaviour of a physical circuit, mathematical methods are used to map specified and physical circuit functionality. Exact matching of the mapping confirms the design with the specification. Imperfect matching can reliably indicate shortfalls or excesses between the specification and the design. It should be noted that the static analysis tools available for software verification, such as SPADE and MALPAS, may also be used to prove the functional correctness of hardware.

5.3.2 Hardware simulation aids

The requirement for simulating ever larger circuits demands ever more CPU time. Fault simulation, for example, can be an Order(n^3) problem, where n is the number of circuit components, so demanding many days of CPU time. There are several aids now available to meet this problem.

Simulation of complex printed circuit boards containing microprocessors and their interfacing circuits creates several problems not least of which is modelling these satisfactorily to a reasonable accuracy. For digital circuits, the modelling requirement is complicated by the lack of gate level representational information for circuit devices. The problem is solved by enabling the actual circuit device in question to be interfaced to a simulation of the rest of the circuit. This obviously removes the modelling problem and allows for an improvement in simulation speed. This method of simulation does not adequately support fault simulation.

Hardware accelerators are also available. In this case the simulation algorithms are speeded-up either by implementation in hardware, or by using parallel processing. The resultant speed increase is claimed to be 2, 3 or even 4 orders of magnitude over that normally obtained on a conventional computer. Benchmarking is required to determine the actual speed advantage in any given simulation. Some accelerators are available for fault simulation. Simulation aids for the hardware in analogue circuits require further development to approach the usefulness of those available for digital circuits.

5.4 Tools to Support Testing

5.4.1 Test pattern generation

Much of the development of testing strategy and testability tools is based upon certain assumptions that have been in use for many years and will, no doubt, continue to be used for many more years. This is despite the fact that new fault classes are appearing that are undetectable by tools based upon the original assumptions. The reason for continuing in this way is primarily because of the computational complexity of testing. Even with the present assumptions, the computer time involved in generating test vectors and verifying fault coverage by fault simulation can often take days of computer time, and sometimes longer, to complete.

Some of the basic assumptions used in testing are:

(a)   Only one fault will be present in a circuit or sub-module for which an independent set of test vectors is being generated.

For digital circuits, the reason for this restriction is easy to explain. If it is required to generate tests explicitly for all multiple s-a-1 and s-a-0 fault occurrences, then the number of faults to be considered is 3^k+1, where k is the number of signal nodes in the circuit. Thus for digital circuits of reasonable size, this quickly becomes expensive in the use of computer time.

Although in practice a multiple fault occurrence might be considered more likely it is still considered reasonable to assume the single fault situation. Practical experience has shown that multiple faults are largely covered one at a time by tests generated to cover all single s-a-1 and s-a-0 faults.

(b)   It is often assumed that fault occurrences in a digital circuit will be of the 'stuck-at' type. It is assumed here that any node in the circuit will only fail by being permanently stuck at logic level 1 or 0 no matter what logic level is being applied to that node.

Other fault types are possible, for example, bridging or stuck-open. Again the 'stuck-at' assumption is still maintained because of the computational complexity of assuming otherwise, and the fact that experience still shows that the other non stuck-at fault types are well covered by tests generated with this assumption.

(c)   For analogue circuits, the same assumption of only one fault being present is made when a diagnostic test sequence is being used. To do otherwise would be too complex to compute in the current state-of-the-art of the less developed testability assessment and verification tools that are being considered for analogue circuits.

It may be possible to further the development of test pattern generators for analogue circuits by using the current state-of-the-art tools for digital circuits by defining signal values for analogue circuits which are equivalent to the 'stuck' states for digital circuits, such as:

For digital	**For analogue**
Stuck-open	no signal open circuit
Stuck at-1	high signal beyond high level limit
Stuck-at-0	low signal below low level limit
Stuck-adjacent	no signal short circuit

It must be emphasised that concern has been voiced as to the practicalities of such a restricted fault model, some fault types being undetectable with tests generated on the basis of the 'stuck-at' assumption.

Test pattern generation tools

Test pattern generation software has the aim of generating a minimal or near-minimal set of test vectors in order to achieve rapid detection and resolution of hypothesised faults.

Most modern test pattern generation tools attempt to generate a specific test for a specific fault, although there are exceptions to this rule. This is achieved by selecting a fault for test generation, such as node XXX being stuck-at logic 0/1 (s-a-0/1), and then (a) attempting to set a logic 1/0 on that node, and (b), creating circuit conditions such that the effects of that fault can be observed at the circuit outputs or some other test observation point. These requirements are based around the concept of path sensitisation which all the test generation algorithms attempt to achieve by various means.

Perhaps one of the best known of these algorithms is Roth's D-Algorithm. This is an algorithm that not only achieves the above, but also guarantees to find a test for a fault should such a test exist. This is achieved by an exhaustive search process. The D-Algorithm was extended and thereby made more efficient by the introduction of some heuristics to remove random decision making in achieving some of its objectives. This algorithm is known as PODEM, which is more efficient in terms of CPU time usage. PODEM and some of its derivatives are in common use for scan designed circuits.

Other algorithms in use include the Critical Path. This is the algorithm often to be found in commercially available packages. It does not guarantee to find a test for a fault even if one exists and so can give incomplete results. The reasons for using this algorithm in preference to the D-Algorithm is that it can achieve an acceptably high fault coverage using less CPU time than is needed for the D algorithm.

All these algorithms will attempt to generate a test for a specific fault, but in doing so, will invariably create conditions for the detection of other faults. Detection of these other faults requires interaction with a fault simulator, thereby making the test generation process more efficient.

Sequential Circuits

Sequential circuits present test pattern generation algorithms with added difficulties. The algorithms can handle combinational circuits relatively easily, but the extensions required for sequential circuits complicates the process. Combinational circuits are straightforward because they contain no stored state devices and no feedback loops. The circuit steady state is purely dependent upon the current state of the inputs. Sequential circuits, however, do contain stored state elements and global feedback loops. They also exhibit a memory condition whereby the present output is dependent upon not only the present input state but also the previous functional history of the circuit. Thus testing for a given fault in a sequential circuit will usually require a sequence of several test vectors before the fault effects are propagated to a circuit observation point. The number of tests required ultimately will also depend upon the ease of initialising the circuit into a known starting state.

The test generation algorithms used are generally iterative extensions of those used for combinational circuits. In many practical cases, test generation for complex sequential circuits is done manually with no recourse to software tools. New semi-automatic software tools are becoming available that are knowledge-based and designed to be an aid to the test generation process.

Random Test Vectors

The use of randomly generated test vectors has gained considerable interest recently, particularly with the development of self-testing circuitry. The vectors may be easily generated using simple hardware in the form of Linear Feedback Shift Registers (LFSRs) also known as Pseudo Random Number Generators. LFSRs will generate a new pseudo-random test vector at every clock cycle. As the test patterns are no longer deterministic, rather more pseudo-random test patterns are required to achieve a given fault coverage than would otherwise be the case. Their popularity is due to the ability to generate pseudo-random patterns simply, at rated circuit speeds in close proximity to the particular circuit area under test. These factors must be weighed against the possible need to perform lengthy fault simulation to verify the fault coverage. Some circuits are more amenable to being tested by random test vectors than others, a fact which might not become apparent until late in the circuit design. It is not easy to predict the number of random test vectors that will be required for the effective testing of a given circuit to a given level of fault coverage.

Testability Assessment Tools

Given the complexity of test pattern generation, its lack of effectiveness on unstructured sequential circuits, and the large CPU time required, an efficient tool for analysing the circuit to determine its likely testability before generating test patterns would appear to have its merits. Testability assessment can be used to identify those areas of the circuit which are, on a relative basis, more difficult to test. These areas can then be modified to improve their testability rating. This is the aim of the many testability assessment tools which are available.

Many of the testability assessment tools base their analysis on the derivation of numerical values for the controllability and observability of nodes in the circuit. Both these factors are important in test generation as they define the ease with which a test vector can apply the required value to a given circuit node, and the ease with which a path can be sensitised from that fault site to the output for fault observation. The numerical values that are produced can also be used to aid in test point selection for either observation or for test signal injection. Having either added test points or redesigned the low testability areas of the circuit, the testability assessment tool may be used again to quantify the improvement in the testability rating of the circuit.

Given that there can be no absolute value for defining controllability and observability, comparison of these values between different circuits has no significance. Comparisons are only meaningful for nodes within the same circuit when changes are made to improve the testability rating.

The main value of these tools is generally considered to be in training designers. Experienced designers/test engineers may be more effective in their assessment of the testability of a circuit and in identifying those areas which are more difficult to test. Nevertheless, some contract authorities may require the use of formal methods for analysing testability during the design phase.

Unfortunately, these tools suffer from the fact that testability can be considered subjective—there is no rigorous definition of testability that can then be applied in some algorithm. Thus the usefulness of these tools is considered by many in the test industry to be over-rated.

5.5 Designing for Testability

Techniques exist for designing circuits such that testing costs are reduced to a minimum. These techniques, known as design-for-testability techniques, have been primarily developed for digital logic circuits, but the partitioning and ad-hoc techniques discussed below can also be applied to analogue and composite analogue/digital circuits.

5.5.1 Partitioning

The development of tests for an electronic circuit can be an extremely expensive task. For purely digital circuits, for example, it is widely accepted that **test development costs vary as the square of the number of logic elements** in the circuits so that doubling the number of elements will result in quadrupling the test development costs.

One way of reducing these high costs is to allow the circuit to be partitioned into a number of modules for testing, the test for each module being generated independently. The problem is then to determine the dividing lines between the modules and define the interfaces between interacting modules.

The best approach to partitioning is to allow **division into functional units**. As an example, a typical microprocessor board will contain functions such as:

> microprocessor
>
> random-access memory (RAM)
>
> program store (ROM)
>
> communications ports
>
> "custom" peripherals

Design-for-testability should ensure that each of these functions can be tested independently. For example, it must be possible to disable the microprocessor so that the other modules can be tested without interference.

Equally, the address decoder must be disabled to allow the microprocessor itself to be tested without involving the other modules. The address and data buses, which are major communications paths between the various functional modules, must be accessed and controlled by the test system.

Partitioning is also crucial when dealing with **circuits containing mixed technologies** such as:

> analogue and digital
>
> TTL and ECL

For example; a typical analogue/digital system may contain functions such as:

> sensor injection (analogue)
>
> sensor signal translation (analogue/digital)
>
> data processing (digital)
>
> actuator signal translation (digital/analogue)
>
> actuator response (analogue).

The transfer functions of these system processes have to be tested. Ideally, the testing should be accomplished without disturbing the system connections. Design-for-testability techniques can assist in creating a design in which transfer functions can be verified without disconnecting the system.

The boundaries between circuit modules should be drawn along or close to the technology boundaries so that the test techniques most appropriate to each technology can be used to verify the transfer function of the various circuit modules.

5.5.2 Ad-hoc design rules

Design-for-testability rules are sets of simple axioms—do's and don'ts—which explain how to produce a more testable design. The rules have evolved over many years in response to the various types of circuit designs and constructional techniques which test engineers have found to give problems during test development, test application, or fault diagnosis.

Design-for-testability rules are applicable to all types of circuit, although their effect on testing costs diminishes as the complexity of the circuit increases. They address issues such as the need to ensure that all stored-state devices in a circuit can be initialised, or the need to provide sufficient space between components to permit access for test probes. Rules relating to both the circuit design and its fabrication are included in these Guidelines.

A useful selection of basic rules is included in Section 7. More detailed guidance can be obtained from many sources, including test equipment vendors and texts on test technology. A list of the more relevant papers and text books on the subject is included in the bibliography.

5.5.3 Scan design

Integrated circuit designers are increasingly adopting rigid design disciplines to solve testability problems. For digital circuits, the most common of these is the Scan design technique.

The concept of Scan design can be explained with reference to figure 1. This shows the Huffman model of a general stored-state circuit. The schematic of many synchronous circuits can be redrawn in this form.

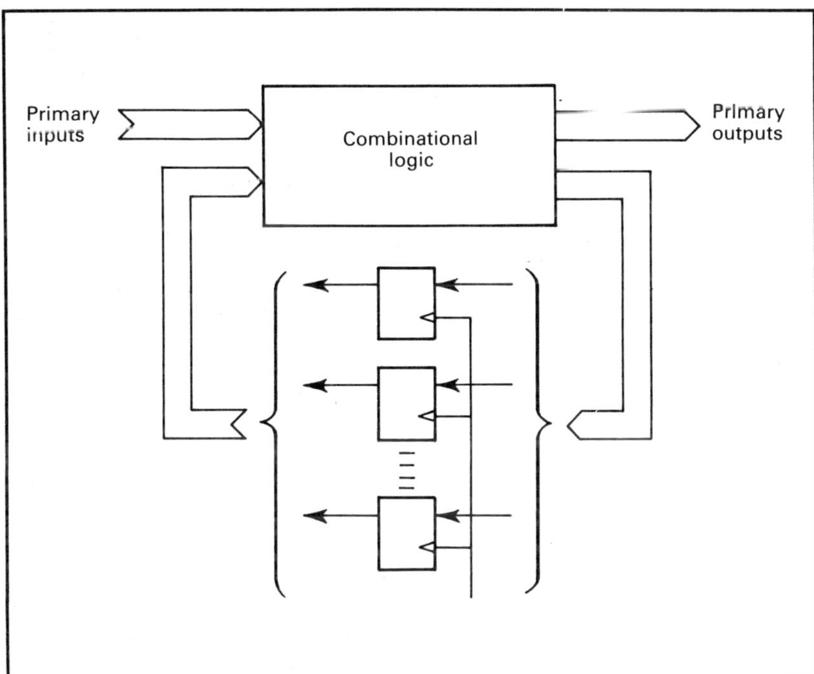

Figure 1—Huffman Model for stored-state circuit

The diagram clearly divides the circuit into two parts. At the bottom is an array of latches which holds the current state of the circuit. It is their presence which causes difficulties in test generation. At the top is a combinational logic block which generates the circuit outputs and the inputs to the latch array from the circuit inputs and latch outputs. We know that test development for this combinational block in isolation would be straightforward.

In order to remove the problems caused by the latch array, we would like to be able to set each latch to an arbitrary state 'directly'—without having to apply lengthy sequences of inputs to the circuit. This would give perfect controllability. Similarly, we would like to be able to see the signals fed to the latches 'directly'—without having to apply lengthy sequences to propagate the signals to the circuit outputs. This would be perfect observability.

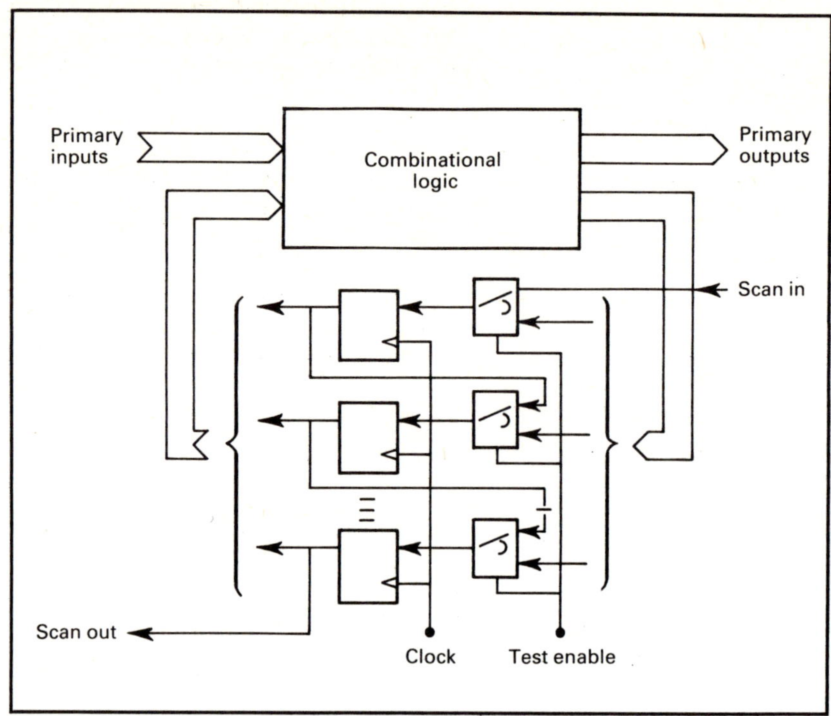

Figure 2—Scan equivalent of Huffman Model

Figure 2 shows a means of changing the circuit so that these two goals can be achieved. Multiplexers have been added at each latch input so that the stored-state device array can operate in two modes:

> **Normal Mode**—where the latches behave as in figure 1. The inputs from the combinational logic are loaded when the clock signal is applied.

> **Shift Mode**—where each latch loads its next state from the previous latch in the chain when the clock is applied.

It is the shift mode of operation which simplifies test development, because it allows latches to be set as required by shifting signals into the circuit through Scan In, and the state of each latch to be examined by shifting signals out of the circuit through Scan Out. Tests can therefore be generated using ATPG packages for the combinational logic block and applied using the shift-register access path.

The Scan design technique, which was first proposed in 1973, has become a key feature of the computer products supplied by a number of large electronics companies. The best known example is from IBM, using a variant of the general technique called Level Sensitive Scan Design (LSSD). In LSSD the multiplexer and latch shown in figure 2 are combined in a single raceless building block called a Shift Register Latch (SRL). Many vendors of application-specific integrated circuits offer one or other of the forms of scan latch in their cell libraries, thereby allowing wider access to the technology.

Designing Scan circuits is straightforward provided that a number of basic design rules are followed. Large companies may have automatic design rule checkers to verify that these rules have been followed. Similar tools can be obtained from some vendors of CAE tools. Test generation tools tailored to the Scan technique are also commercially available. For example, PODEM, mentioned earlier, was developed explicitly by IBM for use within the LSSD environment.

5.5.4 Built-in self-test

Built-In Self-Test (BIST) is a means of reducing the load on external test equipment by incorporating most of the test functions into the circuit itself. The signal generation and comparison facilities of the test equipment are incorporated into the operational circuit design. The functional requirements of the external test equipment may then be substantially reduced.

For example the test equipment may only need to be capable of supplying power and clock signals and receiving the pass/fail indication.

The load on the external support test equipment can only be reduced if the BIST is an integral part of the field replaceable unit. If the BIST is not integral with the field replaceable unit, the external test equipment must be programmed to differentiate between a failure within the BIST and a failure within the operational circuits which the BIST is monitoring.

Because the circuit design disciplines and the construction techniques for BIST are usually similar to those used for the operational circuits which the BIST is monitoring, it cannot be assumed that BIST is functioning correctly at all times. Therefore BIST must be totally built-in as an integral part of the operational circuit that it is monitoring. This can easily be achieved in integrated circuits if BIST is considered at the initiation of a design.

Self-test is aimed at solving the following adverse trends on test costs:

(a) The generation of tests is difficult for complex circuits. Costs have risen at a near-exponential rate as the complexity of circuits has increased. Automatic methods which were adequate in the early 1970s for generating tests can now only be guaranteed to succeed if severe restrictions are imposed upon the design process.

(b) The number of terminals through which the circuit can be accessed during testing is reducing as integration proceeds. Because packaging is expensive, the incentive is always to implement the design with as few external connections as possible. For integrated circuits, the ratio of the number of device pins to the number of logic gates in the device has reduced. More logic has to be tested with less access to the circuit, thereby making techniques such as Scan design necessary to improve controllability and observability. In extreme cases, chips may have internal buses which are not externally accessible.

(c) The operating speed of integrated circuits is steadily increasing as a by-product of integration and it has now reached the stage where it is beyond that achievable by many ATEs. However, advantage of this increased speed is taken in building more powerful computer systems. To ensure that these systems function correctly, testing must be performed as close to full operating speed as possible.

The aim of Built-In Self-Test is to take advantage of the relatively low cost of additional circuitry (particularly that within a VLSI chip) and to incorporate some, or all, of the functions of an ATE within the circuit itself. Thus an operational circuit will contain additional test circuits to capture responses to test stimuli or to generate test stimuli, or both. Because the test circuitry is "built in" the problem of limited external access is removed. Also, because the test circuitry is a part of the total BIST design, it will operate at the same speeds as the circuit being monitored. BIST is likely to be the dominant test technique for integrated circuits in the 1990s and techniques for producing self-testing circuits are still evolving. Some larger companies have tools which are able to convert designs into self-testing equivalents, but these tools are not available commercially.

By building-in the necessary test circuitry then, on failure, it can be removed with the operational circuit that it monitors. Thus the ambiguity of whether a malfunction is in an operational circuit or a built-in test circuit does not affect the repair action. The correct functional performance of built-in test circuits cannot readily be verified by external test equipment, be it either automatic or manual test equipment. Therefore the building-in of test circuitry can allow the use of simpler test equipment for manufacture and maintenance of operational circuits.

5.5.5 Board-specific techniques

A number of design-for-testability options exist which are specific to electronic printed circuit boards.

(a) It is possible to make cheap and reliable contact to the internal connections of the design using a bed-of-nails fixture. Such a fixture allows the state of the internal connections to be observed by the external test equipment and, if overdriving techniques are employed, for direct control of internal connections. Tri-state buffers can be introduced into digital circuits to allow direct control without the requirement for back-driving. Often this can be done at little cost by using a version of a device which has a tri-state output capability in place of one without this capability.

(b) Several integrated circuits are available which provide test support features (see Appendix *B6*). Integrated circuit suppliers are increasingly being encouraged to include facilities in their chips to support board-level product testing. Currently, there is a trend towards the use of boundary-scan techniques, both for military products (e.g. Very High Speed Integrated Circuits—VHSIC) and commercial devices. The boundary-scan path provides shift register access to all chip pins, so allowing many test functions to be performed more effectively (see Section *5.7.2*).

5.6 Test Equipment

Test equipment may be broadly divided into categories which cater for component, circuit board, sub-system and system testing. Within these categories there are equipments which perform either digital or analogue testing only and those which cater for both of these test disciplines. For example, test equipment may be targetted towards the testing of either microprocessors or memory boards and will have different capabilities to suit the special testing requirements of these different types of circuit configurations.

A further division can be made between test equipment which is controlled manually (MTE) and that which is controlled by a stored test program (ATE).

5.6.1 Manual test equipment (MTE)

Manual Test Equipment can be used whenever the manual control times that are introduced into a test sequence by a human operator do not impact the validity of the test results or prolong the implementation of a test sequence beyond an acceptable economic limit.

In general, the scope of the test resources used in MTE can be similar to the test resources that are incorporated into ATE. The main difference is that when using MTE the main flow of testing is controlled by manual actions. Many configurations of MTE are designed to be semi-automatic in that manual initiation of a sub-routine within a test requirement is implemented by an automatically controlled test routine. Usually the read-out of test results is performed manually by a human operator.

When the testing of a UUT demands short testing times, the handling of considerable quantities of test data, or the handling of large numbers of UUT in a timely manner, then there is no alternative to using ATE. The provision of a number of MTE of the same type for the parallel testing of a high throughput of UUT is usually uneconomic compared with using a single fast operating automatic test system.

The proportional cost of testing within the unit cost of manufacturing and maintaining a product is driving both manufacturers and maintainers towards the wider use of ATE throughout the life of a product. Therefore it is appropriate to discuss ATE in more detail.

5.6.2 Automatic test equipment (ATE)

Automatic Test Equipment has evolved to satisfy the requirement to perform accurate and rapid testing of electronic components and assemblies. Characterisation testing, pass/fail testing, or fault finding can be performed, checking functional and parametric capability of digital, analogue, or composite circuits. Test result information can be provided for quality control or factory management systems. Currently the key factors are test speed, pin count, throughput, accuracy, and the ability to be integrated into factory management systems to provide cost effective solutions to the testing problems of complex units.

The choice of ATE in any production environment has a great impact upon the design-for-test strategies that designers will be able to employ effectively.

5.6.3 Basic ATE Organisation

The basic organisation of an ATE is straightforward. A sequence of test stimuli must be applied to the Unit Under Test (UUT) and the results compared with the expected response whilst maintaining signal integrity between the ATE and the UUT. When necessary, the environment around the UUT is maintained within controlled conditions. Automatic handlers are often used with ATE for component testing and signal and environment integrity must be extended to include these interfaces.

Essentially then, ATE comprises:

Test resources for stimulation and response measurement

Memory and instrumentation for stimulus and response evaluation

Instrumentation for parametric measurement

Timing generators to gate input signals and measurement strobes

A high integrity switching matrix which permits a pin on the UUT to be connected to the appropriate input or output circuitry in the ATE

Power supplies which provide power to the UUT

Power supplies which provide reference levels for the instrumentation

A controller which controls and monitors the operation of the total system when executing a test program.

A test system may be a stand-alone equipment or integrated into a factory management system. In an integrated environment a computerised system controller interfaces the test system to a

CAD/CAM/CAT system so that test data can be provided for a CAM system. The system controller may control one or more test stations according to the throughput requirement. For high throughput, a UUT may be loaded via an automatic handler into a 'personality card' which interfaces the UUT to the tester, whilst providing specific load conditions and maintaining signal integrity. For testing integrated circuits, input stimuli and output response signals are routed through a pin electronics matrix under the control of the test station controller. The input stimuli may be preset to a choice of, say, TTL or CMOS. At extra cost, an ATE may incorporate software programmable drivers whose rise and fall times and logic levels may be set individually. Similarly, the response measurement thresholds may be individually set by software control. Such facilities require more reference supplies and a more complex switching matrix involving an increase in both cost and programming time.

5.6.4 ATE architecture

The organisation of an ATE suited to digital component testing has been briefly described. Other test problems require variations in this architecture to achieve cost savings or to meet special needs. For example, one channel in the ATE can accommodate two device pins provided that each pin is dedicated to either input or output. A second example is an in-circuit tester which accesses all component pins and edge pins on a PCB, and tests each component individually by forcing input pins into the required test state. Whilst many connections are made to the printed circuit board via a bed-of-nails, only those ATE channels connected to the UUT need to be active at any time, with perhaps a small extra number required to condition other local circuitry from affecting the test. Thus the ATE manufacturer can reduce ATE costs by providing a few 'fast' channels and many 'slow' channels, thereby ensuring that fast channels are switched to each UUT in turn, as required.

An in-circuit tester applies test patterns to each component individually by forcing inputs to the desired logic value and monitoring outputs. It should be noted that this forcing technique can take the output circuits of those components which are connected to the input pins of the component under test into conditions which are not specified by the semi-conductor manufacturers. This is called "backdriving". In the worst case, prolonged backdriving may result in changes to the characteristics of this circuitry. Whilst experience is that in appropriately controlled backdriving conditions reliability may not be significantly affected, definitive studies into backdriving are, as yet, incomplete.

Considerable test programming effort is needed to provide the necessary test stimuli and response requirements. Complex devices such as microprocessors and their interfacing circuits must be thoroughly understood by the test program writer, and an intimate knowledge of the ATE is also required. High level languages are available to assist in this process and may include translators from the microprocessor code to the ATE code to facilitate these tests. Care must also be taken to design personality cards which permit accurate parametric measurements if the capability of the ATE is to be exploited to the full.

Boards which use scan path techniques require extra hardware and software tools. The long test sequences which have to be shifted into and out of the scan paths need additional memory on the ATE channels connected to these paths. Software which organises the test patterns to suit the scan path organisation is also required.

5.7 Future Trends

5.7.1 Integrated circuits

As circuit complexities increase test costs continue to rise. Also, there is increased competition which means that timescales are reducing as product complexity increases. Because of this, industry is moving towards the use of structured design for test techniques such as scan and self-test, particularly in integrated circuit designs. The use of these techniques, although expensive in terms of circuit size, reduces test costs and makes test development timescales predictable and acceptable.

Several vendors of custom designed or Application-Specific Integrated Circuits (ASICs) now offer design-for-test facilities ranging from basic scan latch cells with which designers can construct scan circuits to systems which enable the designer to create circuits with self-testing as a built-in feature. More vendors will be following this trend.

5.7.2 Printed circuit boards

The need for structured design-for-test approaches has also been recognised for the testing of printed circuit boards and sub-systems. Attempts are currently being made by the Joint Test Action Group (JTAG—see IEE Computer-Aided Engineering Journal, August 1986) to get standardised features to support board testing built into all integrated circuit designs. Without these features, board testing problems will continue to escalate as circuit complexity grows. There is also an initiative in the USA to define a standard Testability Bus for use at printed circuit board test level. This specification is currently being developed by a Working Group of the Test Technology Committee of the Computer Society of the IEEE.

5.7.3 Test equipment

Looking forward, it will not be long before the UUT will incorporate parallel computer processing in order to achieve ever higher rates of data processing than can be provided by a single CPU. This development in operational equipment will demand parallel processing in an ATE controller to cope with the processing speed of the UUT.

Some of the parallel processing in an ATE will be more extensively performed within the 'smart' test resources including the signal switching test resource which communicates with the UUT connection interface.

The further exploitation of circuit integration will reduce the size of test equipment as it is already doing for operational equipment.

This trend towards using more parallel processing in the UUT will drive MTE towards having more automatic control in the execution of UUT test requirements.

The number of test situations which can be economically satisfied by using MTE will be fewer than hitherto. There will be a movement towards the wider use of ATE because it will not be economical to do otherwise for both manufacturers and maintainers. It will be the only way to restrict the rising costs of testing which are associated with the increase in circuit complexity within the UUT whether it be a complete system or a Field Replaceable Unit within a system.

Section Six: Electrical and Electromechanical Products and Complex Systems

6.1 Introduction

This section covers a wide range of equipment that is diverse in both product type and scale of manufacture. It covers a range, for example, from simple solenoid actuators to electric motors, high voltage transformers, megawatt generator sets, robotic cells, electric locomotives, industrial control systems and complex systems for aerospace products and military equipment.

The reader is also referred to the content of Sections 4 and 5, much of which is also relevant to these products.

The topics covered in the section are:

  6.2 General considerations

  6.3 Managing systems testability

  6.4 Testing

  6.5 Test techniques

6.2 General Considerations

6.2.1 Specifying the requirements

In some older established areas of electrical engineering (such as power generation and transport), project specifications may have tended to be based largely on commonly understood working practices. Now that a much wider spectrum of technologies is being incorporated in such products, a common background of understanding can no longer be assumed and a more rigorous approach to specifications needs to be taken.

As mentioned in earlier sections, the correctness and completeness of the initial Requirements Specification for a product is of prime importance However, this is not always easy to achieve. The customer, if known, may not be in a position to predict or specify clearly his usage, or the load and environment to which the product will be subjected. In other cases the customer may not be known, and yet the final location and strategic importance of the installation can have a direct bearing on the design and the testability and maintenance requirements (e.g. power generating or computing equipment on a North Sea oil rig). In such cases, therefore the **design specification must contain a clear statement on the limiting conditions of use.**

The location of the customer's installation can influence the partitioning of the product into sub-units for ease of delivery and for minimising the cost of holding spares. The strategic or economic importance of the product will influence the provision in the design of self-test (BIT) and redundancy features.

6.2.2 Manufacturing strategy

A Manufacturing Strategy document is an important source of information which the designer needs to assist in the preparation of a testability assurance programme.

It should contain such information as:

- proposed location of manufacture for each in-house operation
- sub-contract manufacturing policy
- component sourcing policy
- tooling capability
- machine tool capability (including computer aided manufacturing)
- power plant requirements
- available in-house test equipment
- external test facilities
- manufacturing standards
- quality assurance requirements
- documentation standards
- documentation control procedures
- test strategy

As the design develops, a manufacturing programme plan should be prepared during the Full Scale Development phase as part of the manufacturing strategy, and regularly updated under rigorous change control procedures.

In some cases, it may need to cover the manufacture of introductory models or batches using temporary manufacturing and testing techniques. This phase may place additional constraints on the design to enable testing to be carried out effectively.

Updated Manufacturing Strategy and Plan documents must always be available to guide the designer and project manager.

6.2.3 Maintenance strategy

The Maintenance Strategy document for a product or system needs to be prepared in collaboration with the designer and project manager at an early stage in the product development, to provide an input to the testability assurance programme. The customer may also be involved in its preparation.

The maintenance strategy should cover such topics as:

- available maintenance skills and techniques
- maintenance sub-contract policy
- repair and part replacement policy
- spares holding policy
- preventive maintenance policy
- maintenance manual standards
- training requirements
- feedback reporting and analysis procedure

During the development phase, the content should be extended to include detailed diagnosis and repair procedures, and estimates of reliability, downtime, mean time to repair, availability. Estimates of the cost of resources required in personnel, equipment, training, documentation and spares can then be developed.

A carefully designed feedback reporting system is essential to allow the designer and manufacturing test personnel to improve the testability of the product by analysis of reported field failures and diagnosis problems.

Updated Maintenance Strategy and Plan documents must always be available to guide the designer and project manager.

6.2.4 Test planning

The plans for testing a product or system should always be developed in parallel with the product development as part of the testability assurance programme. This ensures that the **testability of the design is considered as an integral part of the design process.**

A test strategy should be developed by the designer or project manager, in collaboration with manufacturing and maintenance test planners, so that the strategy is harmonised with the relevant manufacturing and maintenance strategies for the product or class of products.

Comprehensive test specifications and procedures for all stages of manufacture from component and material test to final acceptance testing should then be developed to ensure that, in total, all the requirements of the product specification are met. These specifications need to be continually updated throughout the development period. Their generation should never be left as a job for the manufacturing and maintenance test planners to carry out after handover of the design to the production department. Such an unsatisfactory procedure inevitably leads to late production requests for design modifications to improve testability and maintainability.

The reader's attention is also drawn to the information in Section *4.6* on software test planning, all of which is equally applicable to test planning for hardware products or complex systems.

6.2.5 Sub-unit and material procurement

The buying specification and acceptance test procedure of components or sub-units to be supplied by a subcontractor are matters over which the project manager and designer must exercise control.

In particular, care needs to be taken that:

> The buying specification includes any essential testability features;

> Manufacturers' test schedules and procedures are approved and periodically monitored;

> Incoming goods acceptance test procedures are approved;

> Any changes in buying specification and all test procedures are subject to change control;

> Testability features are not traded out without due management consideration of the overall effects on total system production and maintenance aspects and overall life costs.

The development of the Just-in-Time (JIT) concept for the delivery of sub-contract items, whilst reducing the cost of work-in-progress and minimising storage requirements, does inevitably put pressure on the quality manager to accept sub-standard items in order to maintain production flow. **The project manager will need to exercise strong support for the quality function in order to maintain an effective quality level in such circumstances.**

In some branches of the engineering industry, major sub-units are accepted without retest on the basis of a certification procedure carried out on behalf of the supplier. **Such procedures need regular surveillance, as history has revealed a lapse in standards of certification on occasion.**

6.3 Managing Systems Testability

The basic requirements for assuring testability have been dealt with in Sections *1*, *2* and *3* of the Guidelines. These basic principles apply equally to all types and sizes of product and are further developed in the detailed testability rules to be found in Section *7*.

Products or systems incorporating a complex mixture of engineering techniques do have particular problems which are outlined in the following sub-sections.

Methods of test are also evolving and the designer needs to be aware of these possibilities and make suitable provision in the design to allow their use where appropriate.

6.3.1 Special characteristics

Complex systems such as are found in aerospace, process plants, power generation, ships and weapons systems, exhibit a number of special characteristics:

The high cost of down time

A relatively long in-service life expectancy

Modification and enhancement during life

A difficult environment to design for and to simulate for testing

The possibility of hazardous situations arising due to malfunction or operator error

Prolonged inactive periods

Social, environmental and political implications.

These give rise to specific difficulties in managing testability. For example, in complex systems, management has the task of controlling and co-ordinating a number of separate design, manufacturing and installation activities. These may be separate departments or branches of a single organisation or a number of distinct sub-contractors. In such cases, overall design specification, test and maintenance strategies have to be developed and broken down by negotiation into separate interfacing documents for each design area. Particular attention needs to be paid to the following aspects:

The partitioning of the system into separately testable sub-units which should correspond to the partitioning of the tasks between departments or sub-contractors

The definition of interfaces between separately testable sub-units

The degree to which in-service environments and operating conditions can be simulated during design and manufacture

The matching of widely differing timescales for the development and production of various sub-units

The phasing of system build-up and incremental testing during installation

The possibility that the combination of the hardware with the software can only take place at a late stage of on-site commissioning

Ensuring the unambiguous interpretation of interface and performance specifications by teams with different technical backgrounds (e.g. by using a Test Specification language such as ATLAS)

The incorporation of sensors for condition monitoring (see Section *6.5.4*).

6.3.2 Re-use of components

With systems designed for a long service life, there is a natural tendency for users and designers to:

(a)   extend the use beyond the original design life;

(b)   use a well proven design in an environment different from that for which it was designed.

In the case of (a), an example would be the continued use of Viscount and DC8 aircraft, some now exceeding their original design life by a factor of three. Such extended life can only be justified by a comprehensive programme of regular in-service test and inspection routines.

In the case of (b), testing or simulation of the performance of the original design in the new environment must be carried out systematically.

6.3.3 Modification testing and revalidation

Complex systems are seldom immune from modifications during life to

Meet changed requirements

Improve cost/efficiency of operation

Improve maintainability

Incorporate later technology

Replace obsolescent or unobtainable components.

Advantage should be taken of all such occasions to review testability and maintainability aspects to improve efficiency and reduce costs.

Any temptation to cut costs on complete revalidation of the system following a modification should be weighed very carefully. The extent of the revalidation required should only be determined following a formal review of the technical implications of the modification. The cause of a helicopter crash incurring heavy loss of life was traced to lack of testing following a simple modification to a critical component to improve accessibility for maintenance.

6.3.4 Environment simulation

The provision of adequate simulation of the working environment during both design proving and production testing can be difficult and costly, particularly for complex systems. **Failure to cover this aspect in the past has led to excessive on-site commissioning times, failures in early life and in extreme cases, complete disasters.** The starting point must be a complete understanding and specification of the operational and environmental requirements as background to the derivation of the test and maintenance strategy documents. These documents lay down the testing procedures to be carried out at each stage of design, production, installation, acceptance testing and maintenance.

6.4 Testing

Hardware systems undergo two distinct phases of testing:

Design validation—where the functional characteristics of the product and design specification are checked for correctness and completeness against the original requirements. The design validation process should be carried out as part of the Feasibility Study, Project Definition and Full Scale Development phases.

Product testing—where each manufactured product is checked for correctness and completeness against the product design specification. These tests (or a subset of them) are conducted at various stages during the product's life cycle (manufacture, commissioning, operation).

6.4.1 Test phases

(a) **Design validation/type tests**

In the more traditional branches of electrical and electro-mechanical engineering, it is customary for relatively greater visibility to be given to carrying out, on a prototype design, proving tests which are not to be repeated on production units. This allows extra instrumentation to be built in or applied to the prototype tests to collect detailed performance data which, it is assumed, will also represent the performance of manufactured units. This is in addition to the basic role of type tests, which is to validate the design before becoming fully committed to the manufacturing programme.

(b) **Routine factory tests**

Methods of factory test need regular review, particularly for traditional standard products. By utilising new forms of test equipment, and providing built-in sensors, it may be possible to reduce the testing cost and increase the coverage of the test procedures.

For new designs, a test strategy should be worked out, as part of the overall manufacturing plan (see Sections *2* and *3*). This will show the build-up of tests from the component, material and bought-in equipment levels, through the assembly levels to the final factory acceptance and commissioning tests. The design should then be carried out following the guidance given in Section *2.3.5* with the aim of minimising the total test costs by phasing tests as early as practical in the manufacturing cycle.

(c) **Special factory tests**

The customer may wish to negotiate and witness special factory acceptance tests prior to accepting delivery. **The specification and timing of these tests needs to be agreed as early as possible during contractual negotiations and thereafter maintained under strict change control procedures.**

(d) **Site acceptance tests**

It may not be possible to carry out in the factory before delivery certain tests which relate to the special conditions on site or require connection to other site equipment or plant. In such cases **early agreement must be negotiated between supplier and customer covering detailed test requirements, acceptance criteria and availability and access to site and site plant or equipment.**

6.4.2 Specialised test facilities

Certain types of specialised testing may not be within the capability of the equipment or system manufacturer. In such cases the equipment may need to be submitted to an independent test house or research association.

Examples of such test facilities include environmental testing (climatic, shock, vibration, acceleration), electro-magnetic compatibility testing, short-circuit testing of switchgear, non-destructive testing, suitability for use in explosive atmospheres and certification of conformance to national or international standards.

Such tests must be planned and negotiated early in the programme to ensure that a suitable schedule can be arranged with the Test House.

6.5 Test Techniques

6.5.1 Aids to testing and maintenance

The advent of cheaper and more powerful electronics hardware, including microprocessors, gives rise to numerous possibilities for improving the cost-effectiveness and extending the range and coverage of more traditional testing and maintenance methods.

Examples are:

Built-In Test (BIT) facilities. The use of microprocessors and associated software in electromechanical systems, both domestic and industrial, provides new opportunities for building in self-test routines. The latter may be externally triggered or free running. (See Sections *4.5.3*, *5.5.4* and *7.4*).

Condition monitoring. New and more accurate types of sensor are gradually being made available to extend the possibilities of remotely monitoring the state of plant and equipments during operational service. (See Sections *2.7* and *6.5.4*).

Data collection and analysis. The use of digital data collection, transmission and subsequent analysis by computer techniques provides a powerful feedback mechanism for design correction and improvement of design and test techniques.

Associated intelligence. The use of microprocessor techniques can simplify product and process control by eliminating human "rule-of-thumb" methods and allowing cheaper, safer and more efficient operating procedures to be employed. Where the use of a micro-processor is outside the scope of supply of an electro-mechanical device vendor, a microprocessor may be available in another part of the system to provide the necessary intelligence. (See also Section *2.11* on Expert Systems).

Telemetry. Analogue and digital transmission and concentration techniques can simplify system, plant and process control by allowing all the necessary information to be available at a central console position. When combined with remote condition monitoring and built-in self-test, the full adantages can be realised.

6.5.2 Stress and leakage testing

A number of techniques, such as burn-in and leak testing (hydrostatic or gas-based) at the component level and soak test at equipment level, can be employed to identify and remove components which are sub-standard and likely to contribute to system failure in early life. The conditions for these tests need to be carefully controlled to avoid any risk of damaging good components.

6.5.3 Non-destructive testing (NDT)

A number of quality control techniques have been developed to detect flaws in engineering materials. These may be used in:

selecting materials to specifications

monitoring manufacturing processes

monitoring fabrication techniques

obtaining feedback to improve designs

giving warning of deterioration in service

NDT techniques must be able to detect defects of particular types and sizes with some quantified level of reliability which can be presented in the form of a Probability of Detection (POD) Curve. All materials contain defects; however, it is the significance of such defects, evaluated through a failure/life model, which determines the level of detection capability that NDT is required to perform. Increasingly, NDT is asked to provide data on defect type, including characteristics such as size and orientations. Data of this detail is only provided by a few of the available techniques and remains the subject of current research (See "Review of Progress in QNDE" Vols 1-6).

The techniques which are available for defect detection include:

Optical methods

Radiography (X, gamma and neutron)

Ultrasonics

Magnetic particle detection

Penetrant flaw detection

Eddy currents

Stress wave (acoustic) emission

Thermal

Hydrographic interferometry

Acoustic microscopy

A detailed survey of these methods is given in the OUP Engineering Design Guide 09 entitled Non-Destructive Testing. Reference should also be made to the "Quality Technology Handbook" produced by the NDT Centre, Harwell (see Bibliography).

For many systems, NDT on a finished product is being replaced by in-process quality control using NDT techniques. New NDT techniques are being developed to monitor specific properties of engineering materials during manufacture and in use.

6.5.4 Health and usage monitoring

The availability of cheap, powerful and compact electronic systems and accurate sensors is encouraging the wider application of on-line Health and Usage Monitoring (HUM) or condition monitoring techniques. Mileometers have been a feature of motor-car instrumentation from the earliest models, but usage monitoring in more sophisticated forms can now, with advantage, be built into many products. More recently, in-car instrumentation using a computer system to measure fuel economy has become available on some popular models. Higher up the market one can envisage condition monitoring systems covering many aspects of the efficiency and health of a car's electrical, engine, power drive and braking systems. Such techniques are also being applied to electrical power generating plant and process manufacturing plant. Much wider application can be envisaged as more types of cheap and accurate sensors are developed.

The advantages of such systems include

> The ability to predict impending unserviceability in each line replaceable unit
>
> Notification of induced damage
>
> Detection of overload and transient overload conditions
>
> Monitoring of mechanical wear characteristics
>
> Logging of automatically corrected errors in data handling
>
> Early notification of impending mechanical fatigue failure.

The use of sensors to monitor, for example, vibration, the presence of particles, oil analysis, and metal wear and corrosion and combine and relate the results to thresholds for warning or actions can be effective in

> Reducing design tolerance limits to reduce component size, weight and cost
>
> Improving delivered quality by detecting problems during factory tests
>
> Reducing test costs
>
> Reducing maintenance operational costs
>
> Increasing availability
>
> Providing data for feedback to the design process
>
> Reducing spares inventory.

The advantages of incorporating HUM facilities should be considered as part of the testability assurance programme.

TESTABILITY DESIGN RULES

7.1 General Rules applicable to hardware, software and systems

7.2 Rules for hardware

7.3 Rules for software

7.4 Rules for Built-In Test

7.5 Rules for electromechanical equipment

Section Seven: Testability Design Rules

The following lists of rules are intended for use as a design guide and checklist for designers, project managers and manufacturing and maintenance managers in design reviews and design audits. The lists are not exhaustive and the rules need to be selectively applied according to the nature of the project. They are arranged as follows:

7.1 General rules applicable to hardware, software and systems

7.2 Rules for hardware

7.3 Rules for software

7.4 Rules for Built-In Test

7.5 Rules for electromechanical equipment.

7.1 General Rules Applicable to Hardware, Software and Systems

(a) Consider how testability assurance programmes affect management activities and responsibilities.

(b) Ensure that adequate structure exists for managing testability programmes.

(c) Consider the partitioning strategy to maximise testability.

(d) Ensure that each entity has defined attributes which are capable of verification to the depth required by the design philosophy and the production and maintenance strategies.

(e) Ensure that test access to each entity is sufficient to allow complete performance testing and assessment with the minimum disturbance to the permanent structure of the entity.

(f) Base the testability requirements for the product and its component entities on the design, manufacturing and maintenance strategies.

(g) Determine whether the product and each of its modular parts is to be repaired, modified or discarded on malfunction.

(h) Ensure that each product is a testable entity, i.e. its defined functional performance can be verified by the monitoring of its transfer functions—using either Built-in Test (BIT) or externally provided test facilities.

(i) Ensure that each replacement part is a testable entity.

(j) Partition products into modular parts so that all the transfer functions are specific to each testable entity.

(k) Provide sufficient test access for the evaluation of each transfer function that defines:

functional performance

malfunction performance (e.g. degraded mode)

diagnosis of repair action.

(l)   Design testable entities to facilitate the verification of their transfer functions by test methods which impose minimal effects upon operational life.

(m)   Use unusual or untried test techniques and skills with caution.

(n)   Give consideration to incorporating BIT facilities to verify product serviceability and enable fault diagnosis wherever cost effective.

(o)   If the testability of a product relies upon the use of BIT, ensure that the correct functioning of the BIT can itself be established before using it to indicate serviceability or repair action.

(p)   Avoid test techniques (methods or equipment) which can subject personnel to any hazardous conditions.

(q)   Avoid test techniques (methods or equipment) which degrade the operational readiness, performance, reliability or safety of a product.

(r)   Give consideration to the need to verify the transfer functions over the specified environmental conditions.

(s)   Define failure modes and rate their probability of occurrence during manufacture and in-service life.

(t)   Check the testability design of a product by a failure coverage analysis. This should provide a figure for the proportion of failures that can be detected by the verification of the defined transfer functions using both in-built (BIT) and externally provided test facilities.

(u)   Perform a failure coverage analysis to check the isolation of failures to scheduled replaceable parts. This should be performed in three phases to establish:

> The failure coverage provided by the external monitoring of operational transfer functions using only BIT facilities, with all covers in place
>
> The failure coverage provided by the external monitoring of operational transfer functions using both BIT and any external test facilities, with all covers in place
>
> The failure coverage provided by the externally and internally available monitoring of operational transfer functions and other internally accessible transfer functions for fault diagnosis using BIT and any external test facilities, with covers off.

The results of these analyses will guide the determination of an appropriate balance of in-built and external test facilities needed to achieve the required level of product testability.

(v)   Use failure analysis to provide estimated times for the isolation of each fault for each phase of the analysis.

(w)   Where testing cannot isolate a failure unambiguously to a single replaceable part as initially scheduled, then consideration should be given to repartitioning the design to achieve this objective.

(x)   Plan all manufacturing and in-service test facilities on the basis of an optimal blend of hardware and software to achieve the required level of testability within the constraints that arise due to considerations of economics, size, weight, power consumptions.

(y)   Check that all test equipment skills and facilities demanded are available either in-house or by using external test houses.

(z)   Ensure that a buying specification for sub-contract items or components specifies the acceptance test procedure.

7.2 Rules for Hardware

7.2.1 General

(a)   Define precisely all the necessary transfer functions of each testable entity, including its variability due to:

> Spread of component tolerances encountered during manufacture
>
> Variation of component characteristics during in-service life
>
> Changes in operational loading
>
> Environmental operating conditions.

(b)   Avoid close tolerancing of transfer functions, so far as operational performance will allow.

(c)   Set tolerance levels for sub-assemblies so that any resulting tolerance build-up does not exceed that set for the complete assembly.

(d)   Where product reliability depends on redundancy, ensure that each redundant unit is a testable entity and that the transfer functions of each redundant unit can be determined in isolation.

(e)   Provide Test Point Access by the operational connection interface to a hardware product, by a test connection interface or by a combination of the two.

(f)   Locate test connection interfaces on the exterior of a hardware product or on modules within a product, taking into account any manufacturing, operational or environmental requirements.

(g)   Where the removal of components or printed circuit boards for accessing test points is unavoidable, ensure that the removal may be achieved easily

7.2.2 Electronic hardware

(a)   Provide test pads on the printed circuit which are readily accessible without disturbing electronic components or soldered joints, where additional test points for the temporary connection of a test probe prove unavoidable. Ensure that such test pads are not covered with insulant.

(b)   Adopt a standard pitch (currently 0.1 inch, 0.05 inch or metric equivalents) for connections on a printed circuit board layout. This enables standard bed-of-nails contact jigs to be used for several board designs.

(c)   Avoid the use of adjust-on-test techniques unless considered essential to achieve the toleranced value necessary for a transfer function.

(d) Where adjust-on-test is unavoidable the following criteria should be met:

    (i) All adjust-on-test components must withstand the mechanical stresses imposed upon them by adjustment with the appropriate tool during the testing process.

    (ii) The adjustment characteristic must be sufficiently deterministic to allow easy adjustment of the controlled transfer function to within its assigned tolerance.

    (iii) All adjustable components must be provided with a locking mechanism. This must neither disturb the adjustment when it is being set nor allow the adjustment to change from its original setting except for intentional readjustment.

    (iv) Ready access to adjustable components must be provided at the appropriate assembly stage when adjust-on-test is needed and also on the completed assembly, should adjustment be needed at this stage for final functional testing and maintenance testing during service.

    (v) Overlaps must be avoided between the effects upon a single transfer function where the adjustment of more than one adjust-on-test component is required to determine that the single transfer function is within its assigned toleranced value.

    (vi) The adjustment of adjust-on-test components must be effected without the use of specially designed tools, wherever possible.

    (vii) For the alignment of circuits by the adjust-on-test technique, the adjustable components should be preset to provide an approximate value before testing begins so that minimal time is taken to achieve alignment.

(e) Avoid the use of select-on-test techniques unless considered essential to achieve the toleranced value necessary for a transfer function.

(f) Where select-on-test is unavoidable, the following criteria should be met:

    (i) The select-on-test components must be of suitable quality to ensure the necessary stability of the transfer function that is being adjusted by this technique.

    (ii) The adjustment characteristic of the select-on-test technique should be sufficiently deterministic to allow selection of the correct value for the selected component in the minimum number of attempts at selection.

    (iii) There must be ready access for the insertion and removal of selected components at the appropriate assembly stage when the select-on-test technique is needed.

    (iv) The method of selection from a finite range of values for each selected component must be defined completely.

    (v) The mounting arrangement for each selected component must be compatible with the number of insertions and removals needed during manufacture and subsequently at any time during maintenance testing.

    (vi) For the alignment of circuits by the select-on-test technique, an approximate value for each selected component should be fitted before testing begins so that minimal time is taken to achieve alignment.

(g) For the alignment of circuits by either adjust-on-test or select-on-test components, test point access should remain unchanged through the alignment process so that minimal time is taken to achieve alignment.

(h) Provide access to adjust-on-test and select-on-test components and their test points. This should be accomplished without the need to remove parts from a product to facilitate access at the planned adjustment stage for either manufacture or maintenance testing.

(i) Where circuits are primarily mechanical in design (e.g. microwave), check for completeness by a continuity test.

(j) Avoid the need for using in-process conditioning (burn-in) of components, modules or assembled products in order to achieve the necessary tolerancing of product transfer functions.

(k) Whenever the operational configuration of any part of a product has been modified for testing, subsequent tests must be performed to verify that the product has been reinstated correctly prior to its release for service.

(l) Define fully the mechanical mounting requirements and the connection interfaces needed for testing a product and its component parts, covering:

> Mechanical fixing arrangements including any special orientations
>
> Fixing forces or the torques to be applied to fixings
>
> All test discipline dependent connections
>
> Any special interfacing adaptors needed to make the connections
>
> Any special precautions regarding the test environment that are necessary during testing to protect a product from unwanted effects caused by:
>
>> Electromagnetic fields
>>
>> Electrostatic discharge
>>
>> Temperature
>>
>> Humidity
>>
>> Vibration and acceleration
>>
>> Quality of power supplies including overload protection
>>
>> Contamination by fuels/oils/solvents/dirt
>>
>> Wearout caused by testing.
>
> Any special tools needed for:
>
>> The fixing of a product in a test mount
>>
>> The adjustment of re-settable components
>>
>> The handling of a product during testing.

(m) When the configurations of sub-assemblies are similar in appearance, provide means to preclude the incorrect location of the sub-assemblies.

(n) Where possible, avoid the requirement for special test interfacing adaptors which cannot be used for a wide range of products.

7.2.2.1 Rules for digital circuits

(a) Maximise controllability and observability features

Techniques include:

External control or replacement of on-chip/on-board oscillators

Additional logic to allow the breaking and control of feedback paths

Breaking up of long cascaded sequential logic

Provision of test points within PCBs or routed out from within ICs

Use of multiplexers to re-route low controllability/observability nodes to high controllability/observability nodes

Combinations of multiplexers and shift registers

Scan path techniques to allow separate control and observation of the stored-state devices (reconfigured as a serial-in, serial-out shift register) and the remaining combinational sub-circuits

Signature analysis.

(b) Provide a means for initialising the circuit with minimum effort, using techniques such as:

A master reset or clear line

Access to reset/clear lines normally held high or low

Simple routines for initialising the contents of RAM.

(c) Provide facilities to partition large circuits into smaller separated sub-circuits, using transfer gates or multiplexers.

(d) Avoid asynchronous logic. The ATE is unable to control the flow of data through the circuit and fault simulators may be unable to predict circuit behaviour in the presence of a fault.

(e) Avoid monostables or provide over-ride control either over the width of the pulse or over the output pulse itself.

(f) Avoid logically-redundant circuitry.

(g) Provide documentation containing:

A description of the circuit's behaviour

Timing data, valid signal periods and acceptable tolerances

Logic schematic

Power requirements

Details of custom devices

Any special self-test or fault-tolerant features.

(h) Tie any unused floating logic inputs to a power or ground rail through an appropriate resistor.

(j) Terminate any used or unused outputs from tri-state or open-collector logic with a pull-up resistor to prevent logic values arising which will give inconsistent logic signatures.

(k)   For testing a data transmission bus, make provision in the bus protocol and the bus interfacing hardware for a test facility to control the bus traffic during testing.

(l)   Ensure that the line driving interface hardware of a data transmission bus has sufficient circuit driving capacity for the connection of a test facility in reasonably close proximity to the bus hardware.

(m)   Ensure that the use of node forcing (back-driving) techniques during test does not cause damage to associated circuit components.

7.2.2.2 Rules for analogue circuits

(a)   Provide test access to match the frequency bandwidth of the signals being accessed (e.g. use a coaxial connection for signals above 1MHz).

(b)   Use simple signal waveforms for test stimulations as far as possible. (If complex signal stimulation is used, the interpretation of test results can be difficult).

(c)   For diagnosing faults within circuits having feedback loops, make provision for breaking the feedback loop at a point where the open loop impedance is not unduly sensitive to pick-up of unwanted signals during the testing procedure.

(d)   Where low level signals are connected, ensure that the circuits are not being influenced by unwanted signals during testing. For example:

        Select earthing connections to avoid common-mode signal injection and other conducted or radiated interference;

        Take extra precautions where screening has to be removed for testing access.

(e)   Ensure that test signals and power supplies do not include unwanted content which could cause interference.

(f)   Provide an impedance changing buffering circuit between the tester and any high impedance test point.

(g)   Provide means of isolating cascaded circuits to allow the diagnosis of faults to individual stages.

(h)   Use indirect test access techniques to avoid hazards to personnel from high voltages, high thermal dissipations or dangerous radiation levels (e.g. use pick-up loops, transducers).

(i)   Where transducers are used, incorporate them in the interface adaptor as part of the test equipment.

(j)   Avoid the need for special testing facilities as far as possible. Where unavoidable, ensure that the facilities required are fully specified.

7.2.2.3 Rules applicable to both digital and analogue circuits

(a)   Know the characteristics and limitations of the target tester.

(b)   Buffer edge-sensitive signals from the ATE (with, for example, Schmitt triggers), since ATE performance may be slow relative to circuit requirements.

(c)   Keep analogue and digital circuitry apart for test access purposes.

(d)   Avoid mixed logic technologies: make all primary inputs and outputs TTL-compatible, for example.

(e)   Mount complex devices in sockets so that they can be removed for emulation testing techniques. (Alternatively, allow electrical disconnection, for example, via tri-state capability).

(f)   Limit fan-out to one less than maximum to accommodate the additional impedance load of an ATE probe.

(g)   Reduce diagnostic ambiguity by locating equivalent-fault gates within the same package.

(h)   Lay out a board so as to ease the problems of manual probing (same pin orientation for each device, devices in regular columns and rows).

(i)   Provide an easy means of checking for correct board insertion **before** power is applied (e.g. provide a short-circuit link across unused edge-connector positions).

(j)   Where probing is used for test access, avoid multi-board construction.

(k)   Leave room around devices that are to be physically clipped (with a DIP-clip or SMD-clip).

(l)   Collect on-board test observation points into a single test socket.

(m)   Allow track access where current-sensing guided probing may be required (e.g. to diagnose faults visible only on a bus line).

(n)   For boards to be tested on an in-circuit tester:

> Employ single-sided component mounting, where possible
>
> Position contact pads close to the nodes
>
> If devices are constrained in their functionality (e.g. to a BCD-to-Decimal decoder used as a 3-to-8 decoder), then provide facilities to allow the complete functionality to be tested
>
> Ensure that tri-stateable devices can easily be placed into their passive state.

(o)   Surface-mount devices (SMDs) pose special problems because of the variety of packaging and, in some cases, very limited access. Check that, at least, access can be made available to the key control and observe nodes.

(p)   Take precautions to protect circuits from damage due to electrostatic discharge during storage, handling or testing.

(q)   Ensure that circuit configurations are no more complex than is necessary to achieve the defined transfer functions.

(r)   Mark circuit assemblies clearly to show the type, position and orientation of components and the function of controls.

7.3 Rules for Software

(a)   Ensure that the specification of the software is clear, unambiguous, accurate and complete at the requirement, system design and program levels (see IEE "Guidelines for the Documentation of Software in Industrial Computer Systems").

(b)   Identify the units that will be individually modified, and ensure that each is a testable entity.

(c)   Give preference to using well-proven and properly documented software, even possibly at the expense of performance and facilities.

(d)   Define testing methods and tools to be used, before commencing design.

(e)   Where product reliability depends on redundancy, ensure that each redundant unit is a testable entity.

(f)   Where a unit has separate development and production versions (the former instrumented for debugging and testing during development), ensure that the production version is adequately tested.

(g)   Check the software design carefully and regularly at each phase of the development cycle.

(h)   Ensure that the operational software is a precise copy of that tested and deemed to be correct and remains uncorrupted, using established configuration control procedures.

(i)   Ensure that every basic path through a unit of code is tested by its own item of test data.

(j)   Maintain configuration control of the software ensemble, so that the high-level design and documentation is clearly identifiable for each testable entity.

(k)   Adopt a recognised test implementation procedure (see Section *4.3.1*).

(l)   Ensure that the software design takes full account of hardware failure modes.

(m)   Ensure rigid adherence to a suitable documentation plan for all stages of design, development and testing.

(n)   Define fully the program content of all software resident in Read Only Memory (ROM).

(o)   Make provision to verify the software content and hardware integrity of program data stored in ROM (e.g. a Cyclic Redundancy Count (CRC), logic signature or a sum check with memory content identification).

(p)   For testing a data bus, define fully all legal and illegal instructions (words) containing command, data and status content pertinent to the data transmission bus so that GO and NOGO instructions can be applied.

(q)   Define all default conditions for data bus transmission when assigned bits are not selected within a data bus instruction (word).

(r)   Define fully the self-test program of any embedded computer, also ensuring that provision is made for its initiation from an external test facility.

(s)   Where continued degraded operation is required in the presence of faults, co-ordinate testability design closely with design for reconfiguration and configuration verification.

(t)   Provide visibility for verification of the overall and modular transfer functions of software by the top-down structuring of programs according to their implementation.

(u)   Provide full visibility at high levels of program implementation of all those transfer functions hidden at lower levels of implementation which are specifically called at the high levels.

(v)   Provide full visibility at the lower levels of program implementation of all those transfer functions which are hidden at high levels of implementation.

(w)   Pay due regard, at the application software design stage, to the effect and treatment of out-of-range data and invalid data combinations.

(x)   Structure storage into discrete areas or files and restrict access except via the operating system, to prevent unauthorised access by applications tasks.

(y)   Ensure that documentation procedures are specified to facilitate production and evolution through several rounds of change.

(z)   Ensure that large or complex modules contain internal test features and test points.

(aa)   Design comprehensive monitoring facilities at module interfaces.

(bb)   Make effective use of available test facilities of the operating environment (hardware, firmware, operating system and utility libraries).

(cc)   Provide traps or alternate execution paths for detectable error conditions. Annunciate such errors by indicating the point of error, type, time, date and any other relevant data.

(dd)   Design fault codes to be meaningful, to minimise the need to refer to documentation.

(ee)   Conduct an analysis of potential problems covering all system components and check that the software design response is consistent with the design objectives.

7.4 Rules for Built-In-Test

(a)   Design BIT at the system level to respond in a predictable manner to intermittent failures, considering the balance between maximising safety and minimising BIT alarms.

(b)   Ensure that system application software provides adequate facilities (e.g. interrupt and trap capability) to process errors detected by BIT hardware prior to the loss of error information.

(c)   When estimating memory size requirements for a digital system BIT, make sure that sufficient words are reserved:

        In control memory for the storage of micro-diagnostics and initialisation routines

        In main memory for the storage of error processing routines and confidence tests

        In secondary memory (e.g. disk) for the storage of diagnostic routines.

(d) Make sure that the width of each memory word is sufficient to support BIT requirements:

>   In control memory to achieve controllability of hardware components

>   In main and secondary memory to provide for error detection and error correction techniques.

(e) Assign a sufficient number of memory words to non-alterable memory (e.g. ROM, protected areas) in order to ensure the integrity of critical test routines and data.

(f) Design the BIT at system level to localise failures to a small number of items and to advise the user(s) of the degraded mode options. Where necessary design the BIT to isolate a failure to a lower level than a replaceable item in order to determine and to advise the user of those system functions which are lost and those which remain operational.

(g) Make sure the BIT software includes sufficient hardware error detection capability.

(h) Include in the BIT software a method of saving online test data for the analysis of intermittent failures and operational failures which are non-repeatable in the maintenance environment.

(i) Incorporate in the software, or easily-modified firmware, BIT threshold values which may require changing as a result of operational experience.

(j) Incorporate adequate processing or filtering of BIT sensor data to minimise BIT false alarms.

(k) Ensure that the data provided by BIT is designed for both the system operator and maintainer.

(l) Arrange BIT threshold limits for each parameter to optimise the fault detection/false alarm characteristic.

(m) For any product that incorporates BIT facilities, undertake an analysis of the probable operational reliability of BIT hardware and software to determine a realistic figure for product failure coverage.

(n) Provide means for any BIT function to be exercised under the control of the external test controller to verify its correct functioning.

7.5 Rules for Electromechanical Equipment

(a) Avoid hazards to personnel from mechanical motion and stored energy by using indirect access techniques.

(b) Specify appropriate safety equipment and procedures.

(c) Arrange for mechanical tests and adjustments to be completed before energising the electrical system, as far as possible.

(d) Ensure that documentation covers

>   sequence of operation

>   behaviour of equipment when energised

>   tolerance limits of mechanical movement.

(e) Ensure that direction of motion is specified and checked.

(f) Carry out static electrical tests before allowing motion.

(g) Make provision for checking vibration, bearing noise, brush sparking during tests in motion.

APPENDICES

Appendix A: Glossary

ALU	Arithmetic Logic Unit.
ASIC	Application-Specific Integrated Circuit.
ATE	Automatic Test Equipment.
ATLAS	Abbreviated Test Language for All Systems. Used for preparing test requirement procedures which are independent of the test equipment to be used.
ATPG	Automatic Test Pattern Generation.
back driving	An effect produced by individually exercising logic components whilst interconnected to an adjacent circuit.
BILBO	Built In Logic Block Observation; a BIT technique.
black box testing	Testing based on program requirements and specifications, without regard for the internal structure of the program.
Built-In Self-Test (BIST)	see *Built-In Test (BIT)*
Built-In Test (BIT)	An integral capability of the system or equipment which provides an automated test capability to predict, detect, diagnose or isolate failures. It is also used for logging fault profiles.
Built-In Test Equipment (BITE)	Hardware which is identifiable as performing the built-in test function; a subset of BIT.
CAD	Computer-Aided Design.
CAE	Computer-Aided Engineering.
CAM	Computer-Aided Manufacturing.
CAT	Computer-Aided Testing.
checklist scoring	The derivation of a figure of merit by summing the weighted score for each checklist item.
CMOS	Complementary Metal Oxide Semiconductor (a low power consumption integrated circuit technology).
CPU	The Central Processing Unit of a mainframe computer.
CRC	Cyclic Redundancy Count.
ECL	Emitter Coupled Logic (a high speed semiconductor logic technology).
EFS	see *Expected Fault Spectrum*.

electrical partitioning	The isolation of the block of circuitry under test from circuitry not being tested.
entity	A term used to apply generally to any assembly of parts.
Expected Fault Spectrum (EFS)	The distribution by location and type and probability of occurrence of anticipated failure mechanisms. (see also *FMEA*).
failure latency	The elapsed time between fault occurrence and failure indication.
fault coverage	The ratio of failures detected (by a test program or procedure) to failure population, expressed as a percentage.
fault detection	See *fault coverage.*
fault isolation	See *fault resolution.*
fault isolation time	The elapsed time between the detection and isolation of a fault; a component of repair time.
fault modelling	The use of a software logical model to simulate the behaviour of an item under test in the presence of a fault.
failure	Termination of the ability of a functional unit to perform the required function.
failure mechanism	The physical process causing the failure, e.g. short-circuit, missing component.
failure mode and effects analysis (FMEA)	A reliability technique for assessing the distribution and probability of occurrence of anticipated failure mechanisms (see *EFS*).
failure mode, effects and criticality analysis (FMECA)	The assessment of the criticality of anticipated failures in addition to FMEA.
false call	The reporting of a fault when no defect can be found.
fault	The change in circuit operation caused by the failure.
fault resolution	The degree to which a test program or procedure can isolate a fault within an item.
FET	Field-Effect Transistor (a semiconductor device with high input impedance).
FMEA	See *failure mode and effects analysis.*
FMECA	See *failure mode, effects and criticality analysis.*
feasibility study	The period during which the identification and exploration of alternative solutions or solution concepts to satisfy a validated need takes place.
full scale development phase	The period during which the system and the principal items necessary for its support are designed, fabricated, tested and evaluated.
functional partitioning	The implementation of each function on a single replaceable item.
HUM	Health and Usage Monitoring.
IKBS	Intelligent Knowledge Based System.
interface adapter	A unit providing mechanical and electrical connections and any signal conditioning between the unit under test and the automatic test equipment.

initialisation	In testing, the setting of an entity into a known state. In computer programming, the process of setting the values of a variable to a specified value at the start of a program execution. In computing, the loading into main memory of the initial part of the operating system to enable the system to proceed under its own control.
JIT	Just-In-Time.
LSSD	Level Sensitive Scan Design.
MOS	Metal Oxide Semiconductor.
MSI	Medium Scale Integration (integration in the range of 100 to 10,000 transistors on a single semiconductor chip).
MTE	Manual Test Equipment.
NATLAS	National Testing Laboratory Accreditation Scheme.
off-line testing	The testing of an item when removed from its normal operating environment.
Petri net	A system modelling technique using a specific graphical notation devised by C A Petri.
PCB	Printed Circuit Board.
production and deployment phase	The period from production approval until the delivery and acceptance of the last system.
project definition phase	The period during which candidate solutions are selected and refined.
retest OK	A unit under test that malfunctions in a specific manner during operational testing, but performs that specific function satisfactorily at a higher level maintenance facility.
ROM	Read Only Memory.
SRL	Shift Register Latch. A key constituent of LSSD.
Schmoo plot	A graph showing the behaviour of a particular circuit characteristic as two independent parameters, such as operating temperature and power supply line voltage, are altered.
SMD	Surface Mount Device (a form of component packaging and mounting).
SSI	Small Scale Integration (generally fewer than 100 transistors on a single semiconductor chip).
testability	The inherent property of a testable entity which facilitates the verification of its defined properties and performance. This leads to the diagnosis of malfunctions in terms of its characteristic transfer functions, in a timely and affordable manner.
testability effectiveness tracking	The systematic tracking and evaluation of testability effectiveness throughout the manufacturing and in-service phases.
test effectiveness measures	Examples are fault coverage, fault resolution, fault detection time, fault isolation time and false alarm rate.
Test Exploration Ratio	The proportion of paths covered by a software test set, in white box testing.

test requirements document   A specification containing the required performance characteristics of a UUT and specifying the test conditions, values (and allowable tolerances) of the stimuli, and associated responses needed to indicate correct operation.

TTL   Transistor-Transistor coupled Logic (a widely used series of semiconductor logic technology families).

UUT   Unit Under Test.

VHSIC   Very High Speed Integrated Circuit

VLSI   Very Large Scale Integration

white box testing   Testing which takes into account the internal structure of the program, with the aim of testing every path and executable statement.

Appendix B: Some Available Tools

The following information is by no means intended to be an exhaustive list of available tools, but may serve to indicate the sources for obtaining up-to-date information on availability and usefulness.

B1 Logic Modelling	LOGMOD (Logic Modelling)	Detex Inc
	STAMP (System Testability and Maintenance Program)	Arinc Research Corporation
B2 Testability Checklists	RADC Technical Report TR-79-329 (1980)	Rome Air Development Centre
	MIL-STD-2165 Appendix B (1985)	US Department of Defense
	Testability Advisor	Logical Solutions Technology Inc.
B3 Controllability-Observability and Analysis Tools	DTA (Daisy Testability Analysis)	Daisy Systems Corporation
	COPTR (Controllability-Observability Program Testability Report)	Calma Company
	HITAP Testability Analysis Program	GenRad
B4 Simulation Tools	HILO	GenRad
	LASAR	Teradyne
	CADAT	HHB—Systems
	SPICE (analogue)	Various Vendors
	ASTEC 3 (analogue)	SIA (UK agent)
	Analog Workbench	Analog Design Tools Inc
B5 Test Generation Tools	HITEST	GenRad
	TEST SCAN	Gateway Design
	BITGRADE	Gateway Design
	LASAR 6 (PROSECUTOR)	Teradyne
	CADAT (THESEUS)	HHB Systems
B6 Testability Chip Sets	SMT Testability Chip Set	Logical Solutions Technology Inc
	Serial Shadow Shift Register Chip Set	Advanced Micro Devices (AMD)
B7 Safety Critical Analysis	MALPAS	Rex, Thompson and Partners
	SPADE	Program Validation Ltd

B8 Software Engineering

Readers are advised to consult the guidance publications prepared under the STARTS (Software Tools for Application to large Real-Time Systems) programme run by the National Computing Centre (NCC) and funded by the UK Department of Trade and Industry (DTI). These are the "STARTS Guide", the "STARTS Purchasers' Handbook", and a series of debrief reports. Attention is also drawn to the service provided by the Software Tools Demonstration Centre (STDC), also run by the NCC and funded by the DTI. Further information may be obtained from the STARTS Secretariat, NCC, 11 New Fetter Lane, London EC4A 1PU.

B9 Software Testing

Readers are referred to the listings of tools in the literature. A comprehensive listing is contained in Appendices A and B of "Software Testing and Evaluation" by R A DeMillo et al, published by the Benjamin/Cummings Publishing Company Inc of Menlo Park, California, USA.

Appendix C: Standards

C1 British Standards	BS 5750	Quality Systems
	BS 5760	Reliability of systems, equipments and components
	BS 5887	Code of practice for testing of computer-based systems
	BS 6488	Code of practice for configuration management of computer based systems
C2 UK Ministry of Defence Standards	DEF STAN 00-10	General design and manufacturing requirements for service of electronic equipment
	DEF STAN 00-13/2	Guide to the achievement of testability in electronic and allied equipment
	DEF STAN 00-14/2	Guide for the defence industry in the use of ATLAS
	DEF STAN 00-16/1	Guide to the achievement of quality in software
	DEF STAN 00-17	Modular approach to software construction, operation and test (MASCOT)
	DEF STAN 00-40	Achievement of reliability and maintainability

(See also *NATO standards adopted by the MOD*)

C3 USA Department of Defense Standards	MIL-STD-415D	Test provisions for electronic systems and associated equipment, design criteria for
	MIL-STD-1388	Logistic support analysis
	MIL-STD-1521	Technical reviews and audits for systems, equipments and computer software
	MIL-STD-1629A	Procedures for performing a failure mode and effect analysis
	MIL-STD-2165	Testability program for electronic systems and equipments
C4 IEEE Standards	IEEE Std 416	IEEE Standard ATLAS test language
	IEEE Std 488	IEEE Standard digital interface for programmable instrumentation
	IEEE Std 716	IEEE Standard common ATLAS test language

	IEEE Std 771	Guide to the use of ATLAS
	IEEE Std 828	Standard for software configuration management plans
	IEEE Std 829	Standard for software test documentation
	IEEE Std 982	Software reliability measurement
	IEEE Std 1008	Software unit testing
	IEEE Std 1016	Recommended practice for software design descriptions
C5 NATO Standards	AQAP-1	NATO Requirements for an industrial quality control system
	AQAP-2	Guide for the evaluation of a contractor's QC system for compliance with AQAP-1
	AQAP-13	NATO Software quality control system requirements
	AQAP-14	Guide for the evaluation of a contractor's software QC system for compliance with AQAP-13
C6 UK Health and Safety Executive	HSE	Programmable electronic systems in safety-related applications 1. An introductory guide 2. General technical guidelines
C7 UK National Testing Laboratory	NATLAS	Software Unit Test Standard and Method

Appendix D: Bibliography

1. Ambler A P, Baker K and Musgrave T (eds); 'Testability: Advancing VLSI Quality', Prentice-Hall, 1987.

2. Ambler A P, Manning R L and Muhammed N; 'Hardware Accelerators for CAD': IEE Computer-Aided Engineering Journal, August 1985.

3. Arsenault J E and Roberts J A (eds); 'Reliability and Maintainability of Electronic Systems', Computer Science Press, 1980.

4. Bateson J; 'In-Circuit Testing', Van Nostrand Reinhold, 1985.

5. Beizer B; 'Software Testing Techniques', Van Nostrand Reinhold, 1983.

6. Bennetts R G: 'Introduction to Digital Board Testing', Crane-Russak/ Edward Arnold, 1982.

7. Bennetts R G: 'Design of Testable Logic Circuits', Addison-Wesley, 1983.

8. Bennetts R G and Jack M A (eds); 'Design for Testability', IEE Proceedings Vol 132 Part G No 3, June 1985.

9. Berger H (ed); 'Nondestructive Testing Standards—a Review' Special Technical Publication 624, American Society for Testing and Materials (ASTM), 1977.

10. Breuer M A and Friedman A D; 'Diagnosis and Reliable Design of Digital Systems', Pitman, 1976.

11. Burchon D (ed); 'Non-Destructive Testing', Engineering Design Guide 09, OUP (second edition, in preparation).

12. Bussert J; 'Comparing Testability Measurement Tools', Defense Electronics, June 1986.

13. Chandrasekaran B and Radicchi S (eds); 'Computer Program Testing', North Holland, 1981.

14. Davis B; 'Economics of Automatic Testing', McGraw-Hill 1982.

15. DeMillo R A et al; 'Software Testing and Evaluation', Benjamin/ Cummings 1987.

16. De Paul R A; 'Logic Modelling as a Tool for Testability', IEEE Autotestcon Proceedings, 1985.

17. EEUA; 'Guide to the Engineering of Microprocessor Based Systems for Instrumentation and Control', EEUA Handbook No 38, 1981.

18. Fagan M E; 'Design and code inspections to reduce errors in program development', IBM Systems Journal Vol 15, No 3, 1976.

19. Fujiwara H; 'Logic Testing and Design for Testability', MIT Press, 1985.

20. Health and Safety Executive; 'Programmable electronic systems in safety-related applications', HMSO, 1987.

21. Hetzel W; 'The Complete Guide to Software Testing', QED Information Sciences, 1984.

22. IBM; 'Inspections in Applications Development—Introduction and Implementation Guidelines', IBM Publication GC 20-2000-0, 1978.

23. Marsh C D (ed); 'Guidelines for the Documentation of Software in Industrial Computer Systems', IEE, 1985.

24. Maunder C M and Bennetts R G; 'HITEST—Using Knowledge in Test Generation', IEE Computer-Aided Engineering Journal, Juner 1985.

25. McCabe T J (ed); 'Structured Testing', IEEE Computer Society Press, 1983.

26. McCluskey E J; 'Logic Design Principles: with Emphasis on Testable Semi-custom Circuits', Prentice-Hall, 1986.

27. Miczo A; 'Digital Logic Testing and Simulation', Harper and Row, 1986.

28. Morris D S; 'In-circuit, Functional or Emulation—Choosing the Right Test Solution', IEE Computer-Aided Engineering Journal, June 1986.

29. Myers G J; 'The Art of Software Testing', Wiley-Interscience, 1979.

30. NDT Centre, Harwell; 'Quality Technology Handbook' 4th Edition, Butterworths, 1984.

31. National Computing Centre; 'The STARTS Guide', NCC, 1984.

32. Ould M A and Unwin C (eds); 'Testing in Software Development', British Computer Society Monographs in Informatics, Cambridge University Press, 1987.

33. Parker K P; 'Testability: Barriers to Acceptance', IEEE Design and Test, October 1986.

34. Penfold S L and Coupland J W (ed); 'Expert Systems II; a Bibliography on Expert Systems in Electrical and Electronic Engineering', IEE, 1986.

35. Peterson W W and Weldon E J; 'Error Correcting Codes', MIT Press, 1972.

36. Prince G and Shortland M; 'Condition Monitoring—the Automated Manufacturing Perspective', British Journal of NDT, March 1987.

37. Pynn C; 'Strategies for Electronics Test', McGraw-Hill, 1986.

38. Roth J P; 'Computer Logic Testing and Verification', Pitman/Computer Science Press, 1980.

39. Sharp R S; 'Organisation and Range of NDT Research in the UK', British Journal of NDT, Vol 24, pp 303-310, 1982.

40. Strover A C; 'ATE: Automatic Test Equipment', McGraw-Hill, 1984.

41. Thompson D O and Charenti D (eds); 'Review of Progress in QNDE' Vols 1-6, Plenum.

42. Timoc C G (ed); 'Selected Reprints on Logic Design for Testability', IEEE Computer Society Press, 1984.

43. Turino J L; 'A Proposed Standard Testability Bus', Logical Solutions Technology Inc., 310 W. Hamilton Avenue, Campbell, CA 95008, USA.

44. Wilkins B R; 'Testing Digital Circuits: an Introduction', Van Nostrand Reinhold, 1986.

45. Williams T W (ed); 'VLSI Testing', (Advances in CAD for VLSI Vol 5), North Holland, 1986.